ABD EL HEDI GABSI

# The Art of Precision

ABD EL HEDI GABSI

# The Art of Precision

## How CNC machine tools work

ScienciaScripts

**Imprint**

Any brand names and product names mentioned in this book are subject to trademark, brand or patent protection and are trademarks or registered trademarks of their respective holders. The use of brand names, product names, common names, trade names, product descriptions etc. even without a particular marking in this work is in no way to be construed to mean that such names may be regarded as unrestricted in respect of trademark and brand protection legislation and could thus be used by anyone.

Cover image: www.ingimage.com

This book is a translation from the original published under ISBN 978-620-6-70337-2.

Publisher:
Sciencia Scripts
is a trademark of
Dodo Books Indian Ocean Ltd. and OmniScriptum S.R.L publishing group

120 High Road, East Finchley, London, N2 9ED, United Kingdom
Str. Armeneasca 28/1, office 1, Chisinau MD-2012, Republic of Moldova, Europe
Printed at: see last page
**ISBN: 978-620-7-89452-9**

# Contents

Numerically controlled machine tools are increasingly used in mechanical manufacturing companies. CNC manufacturing refers to the machining of mechanical parts using programmable machines. These machines are mainly used to manufacture complex parts with a high degree of precision. Thanks to the integration of cutting-edge technologies, they have undergone a remarkable evolution, becoming more powerful, easier to use and more precise.

There are several programming standards for CNC machines. The first programming standard was ISO 840, developed in 1973 by the International Organisation for Standardisation. Today, these standards have evolved and diversified considerably. This book mainly presents the ISO 4649, DIN 66025, RS274 and FANUC programming languages. In short, these languages are based on codes that determine the operations to be performed by the machine. Although the basic codes are the same for most machines, the codes for auxiliary functions, structures and subroutines differ according to the standards.

An NC program is essentially a list of operations carried out automatically by the machine. It is made up of "G codes" and "M functions". These codes are initially entered into the machine controller, then converted into commands for the pre-actuators, and finally transformed into actions. These actions can be slide movements, spindle rotations, tool changes or coolant commands, etc. All the codes used to produce a part make up the NC program. This program can also contain instructions such as conditions, loops, counters and mathematical operations, etc.

To help manufacturers master CNC manufacturing, this book first describes the basic concepts of CNC machines. Secondly, it focuses on the programming language known as "G codes". Finally, it describes machining cycles to complete the learning process.

Chapter I

# Numerically controlled machine tools

## 1.1 Introduction

Parts manufacturing processes have progressed considerably. Today, numerically controlled machines have replaced conventional machines in the majority of mechanical manufacturing industries. From a functional point of view, a numerically controlled machine tool, or "NCMT", is essentially the same as a conventional machine tool. The difference between a NC machine tool and a conventional machine is in the control section. In conventional machines, operations such as spindle rotation, slide movements, coolant, tool change and speed change are carried out by the operator, whereas in CNC machines, these operations are managed by the machine controller.

## 1.2 Historical background

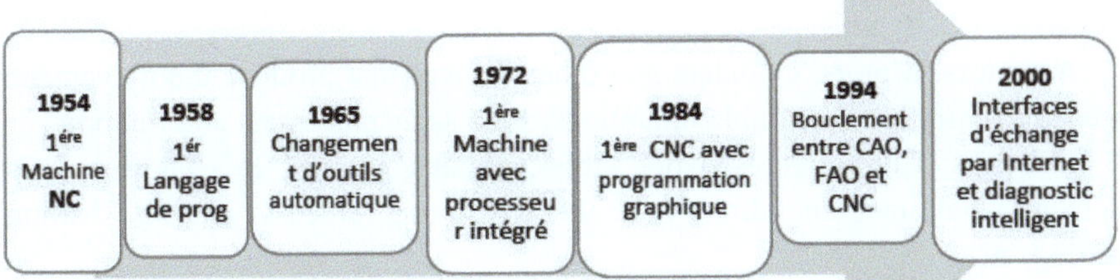


Numerically controlled machine tools" "Numerically controlled machine tools": this term generally refers to mechanical manufacturing machines automated by chip removal, such as milling machines, lathes, electro-erosion machines, etc. However, 'NC' concepts are currently being applied to a wider range of machines, such as cutting machines, punching machines, assembly machines, welding machines, etc. These machines, although using different processes, benefit from the advantages of numerical control, offering greater automation, precision and flexibility in their manufacturing process.

## 1.3 Advantages and disadvantages of NCMs

**Advantages :**

✔ Manufacturing precision: higher quality parts.

✔ Machining complex surfaces.

✔ Flexibility: favours small series production.

✓ Repeatability.

✓ Safety: no human operator.

**Disadvantages :**

✓ Very high initial investment.

✓ Qualified operators.

## 1.4 Operating principle

Like all automated systems, NC Machines are made up of a control part ("PC"), an operating part ("OP") and a linking part made up of pre-actuators and sensors.

The control section converts the programmes into pre-actuator control. It generally comprises a control director, a programmable logic controller (PLC) and electronic cards. In some machines, the PLC can be replaced either by a microcomputer or by an electronic board.

The operating part is used to move or rotate the machine and is made up of two parts:

• Actuators: Motors, cylinders and other devices that produce the movements needed to carry out machining operations, such as displacement and rotation.

• Effectors: Cutting tools, work tables, chucks, pumps, etc. These are the elements that intervene directly in the machining process, cutting, drilling, milling, etc.

The link section plays a crucial role in ensuring communication and dialogue between the control and operating sections. This enables instructions and data from the CNC programme to be transmitted to the actuators and end-effectors, so that machining operations can be carried out automatically and accurately.

## 1.5.1 Architecture

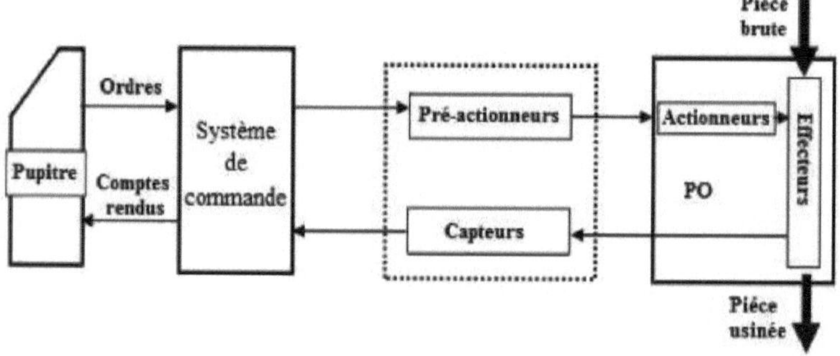

### 1.5.2 Functional structure of a NCM

The functional structure of a numerically controlled machine tool (NCMT) is designed to ensure automated and precise operation of the machining process. It is based on the harmonious integration of the control, operating and connecting parts, enabling advanced automation of the machining process to meet the requirements of quality, precision and efficiency in the manufacturing industry.

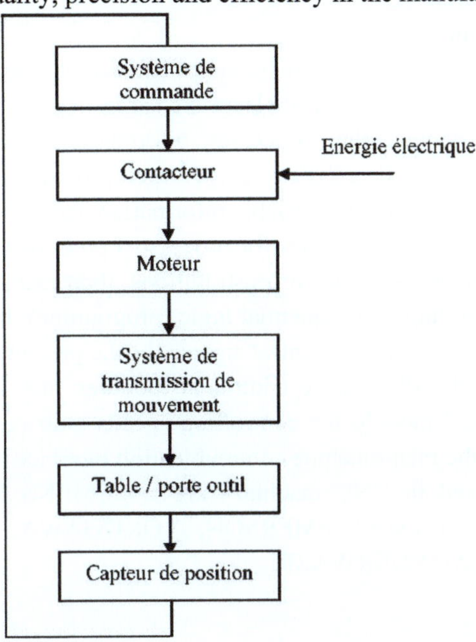


### 1.5.3 Power supply system

This system is located in the electrical cabinet and supplies the machine (motors, pump, console, lamps, etc.) with electrical power. The main components of this system are as follows:

•   Transformer: Its role is to reduce the amplitude of the electric current, thus adapting the voltage to the specific needs of the machine.

•   Stabilised power supply: This part of the system converts alternating current into direct current, guaranteeing a stable power supply for the various machine components.

•   Contactors: Contactors are controlled by the control unit, and their function is to interrupt or authorise the passage of electrical current to the motors and other machine devices, depending on machining requirements.

•   Circuit-breakers: Circuit-breakers play a protective role by interrupting the flow of electrical current in the event of an overload or short-circuit, thus preventing damage to the machine's electrical components.

• Protective elements: These include fuses, thermal relays, surge protectors, etc. Their role is to protect the machine against overvoltages, overcurrents, short circuits and other electrical problems that may arise. Their role is to protect the machine against overvoltages, overcurrents, short circuits and other electrical problems that may arise.

### 1.5.4 Control system

Most CNC machines are controlled by control managers, programmable logic controllers, also known as PLCs, and electronic cards.

The PLC is the element that processes information according to a pre-established programme. It consists of a microprocessor and a memory containing the machine software, variable information and data.

In NC machines, the calculation unit (or processor) processes the data received via the console, sensors, switches or push-buttons, then manages the operation of the actuators according to a sequential logic (programme). It converts the "G" code machining programme into control signals for the pre-actuators.

In CNC machines, the PLCs have additional modules such as the "counting" module, the "motion" module for controlling speeds and movements, and the "HMI" module for the man-machine communication interface.

The main PLCs used in CNC machines are : SIEMENS, OMRON, Allen-Bradley, Schneider, FANUC, EMERSON, YOKOGAWA, INTELLUTION, FOXBORO, ABB, WONDERWARE.

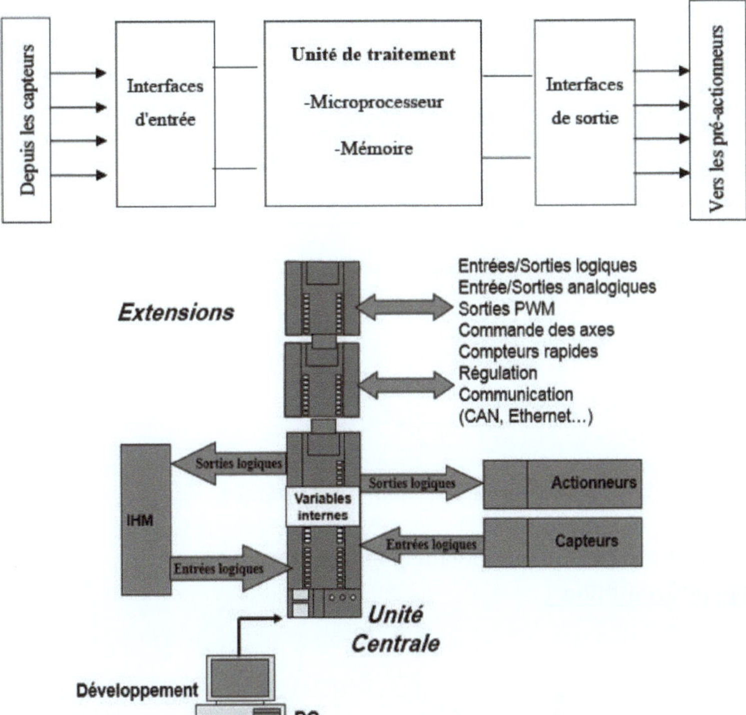


*Figure 1. Architecture of a programmable logic controller*

Initially, the control manager sends a signal to the servomotor, while continuously monitoring the axes through feedback from sensors or encoders. It processes this information in real time to generate control commands based on the CNC programme.

The main functions of an order manager are as follows:

- Interpret the CNC programme.
- Executing moves.
- Process feedback from trips.
- Correcting travel errors.
- Managing data and measurements.

The command director plays a crucial role in the precise control of machine movements, ensuring close coordination between the actions of the servomotors and the commands of the CNC program. This ensures that machining operations are carried out efficiently and accurately, contributing to the quality and reliability of the mechanical manufacturing process.

<u>**Linear interpolations :**</u>

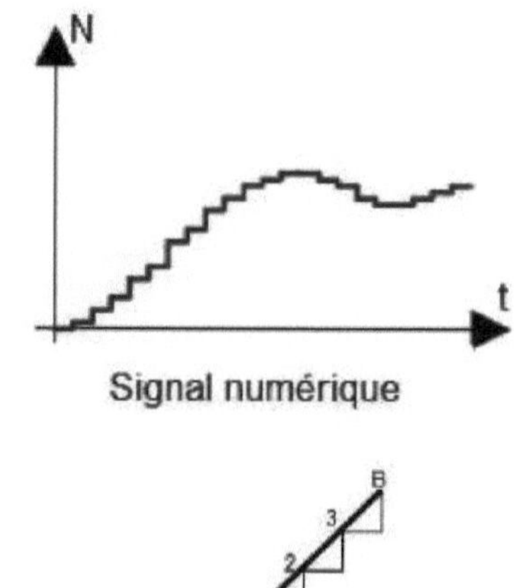



*Figure 2: Linear interpolation*

Displacement is based on interpolation and approximation.

<u>**Circular interpolation :**</u>

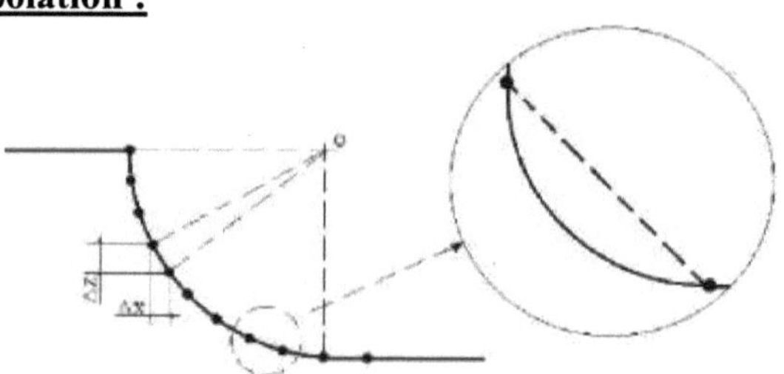

*Figure 3: Circular interpolation*

A circular displacement is the result of combining two linear interpolations.

<u>**Moving process :**</u>

Trolleys are moved in three phases:

- From the initial position to **t1**: the speed gradually increases from "Zero" to maximum speed.

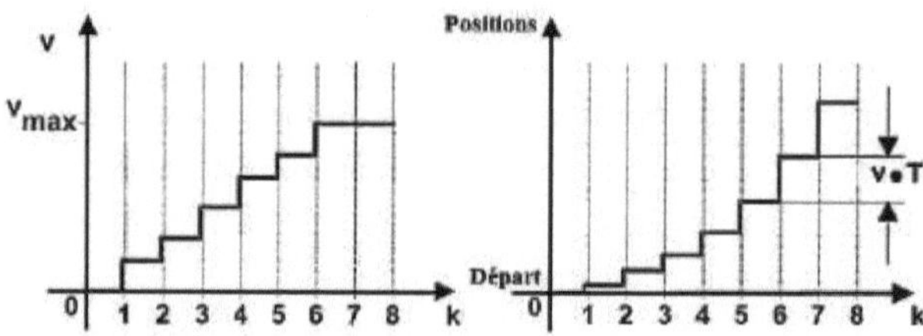

-    From **t1** to **t2**: the carriages move at maximum speed.

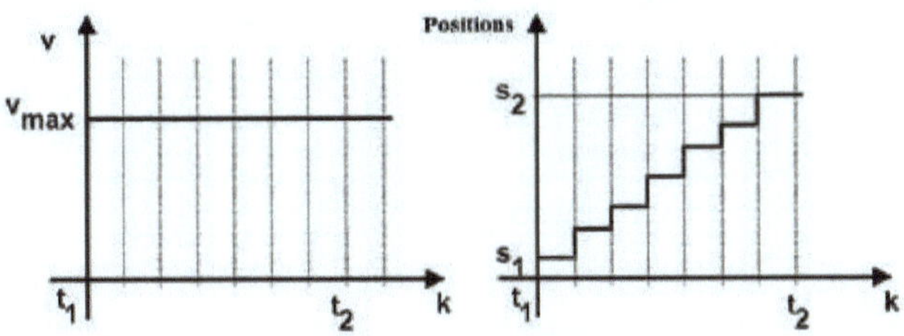


-    From **t2** to the position to be reached (**t3**): the speed decreases until it reaches the "Zero" value.

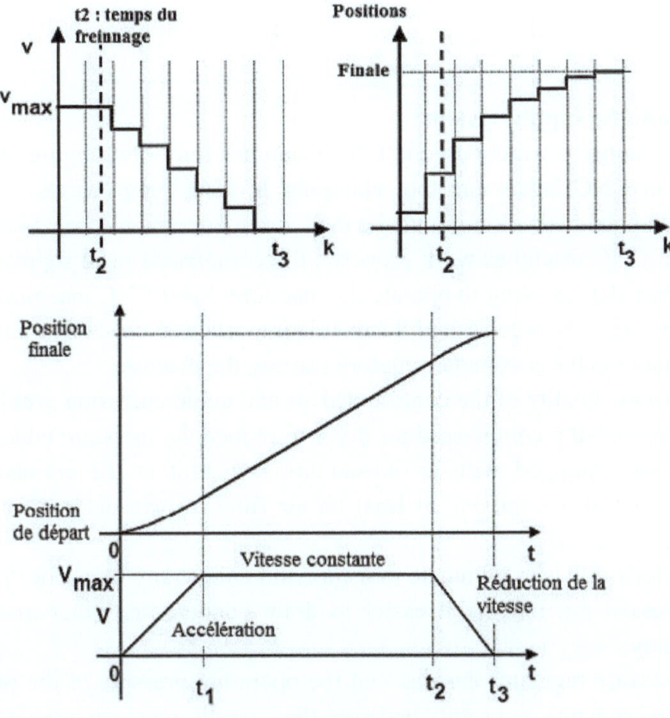


*Figure 4. Movement phases*

## 1.5.5 Buses

Buses are used to transmit and exchange information.

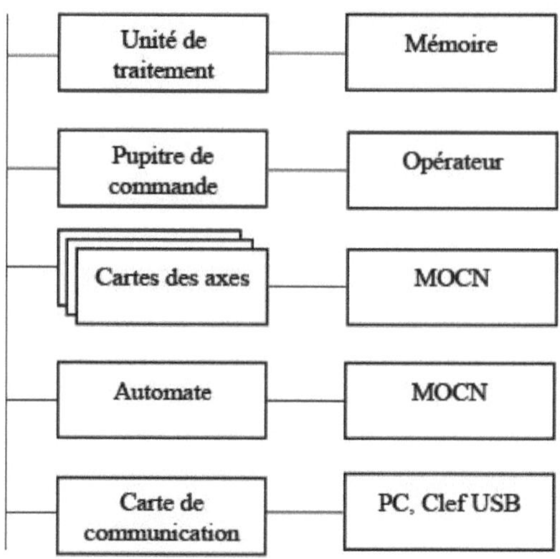



## 1.5.6 Pneumatic supply system

Pneumatic energy is widely used in CNC machines for tool changing, workpiece clamping in CNC lathes and tool clamping in machining centres. For safety reasons, the pneumatic circuit must be kept away from the electrical cabinet. To ensure that CNC machines work properly, the compressor must supply a higher pressure than that required to operate the machine. Most CNC machines operate with a pressure of between 6 and 8 bar. It is important to check that a reserve of air is available in the accumulator before starting the machine.

To preserve the quality of the compressed air and avoid corrosion problems, it is advisable to install a compressed air dryer to reduce the moisture content. CNC machines are equipped with an air-conditioning unit at the entrance to the pneumatic circuit, comprising at least an air filter, a pressure regulator and a lubricator.

- The filter's role is to eliminate moisture and collect any impurities present in the compressed air, making it easier to drain condensate, either manually or automatically.

- The pressure regulator ensures that the operating pressure of the pneumatic components remains constant, because the supply pressure created by the compressor is higher than that required for the operation of CNC machines.

- The air lubricator sprays an oil mist which is transported by compressed air to lubricate cylinders and distributors.

Distributors in CNC machines play an essential role as the link between the PLC and the actuators. They switch the direction of compressed air flow to drive the cylinders and clamps, ensuring precise, efficient operation of movements in the machining process.

### 1.5.7 Lubrication system

Lubrication is a fundamental factor in the smooth operation of machining centres. This system generally consists of an oil reservoir, filter, pump, non-return valve and distribution pipe. Its main role is to reduce friction and wear on moving parts such as screw-nuts, bearings, bushes, slides, etc.

In CNC machines, lubrication is carried out continuously and automatically, ensuring smooth operation and increased durability of mechanical components. Once the components have been lubricated, the oil is returned to the reservoir, where it is filtered to remove any impurities that could harm the lubrication system. This process ensures that the oil remains clean and effective to protect moving parts and ensure high-quality machining.

### 1.5.8 Motion transmission system

This system is essential to ensure precise and efficient movement of the table or carriages in CNC machines. To meet these precision requirements, manufacturers frequently use a ball screw system, renowned for its ability to reduce friction and facilitate the transmission of high power.

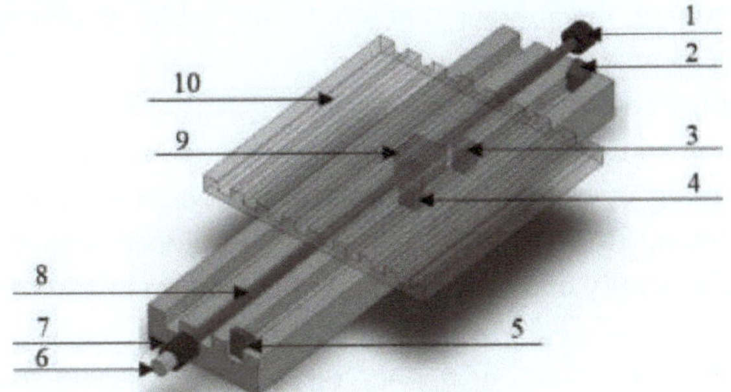


*Figure 5: X-axis control on a CNC machining centre*

1   :   Encoder
2   :   Limit stop
X+
3   X+ limit switch
4   Limit switch X
5   Limit stop X
6   :   Dynamo
tachometric
7   :Electric        motor
8   :Ball       screws
9   :   Nut

10 : Table

The main components of a shaft movement system are as follows:

- A DC or AC electric motor.
- A motion converter that transforms the motor's rotational movement into table or carriage movement (often a ball screw and nut system).
- Two end stops.
- Two sensors.
- A coder.
- A tachometric dynamo that accurately measures rotation speed.
- Motion transmission components such as bearings, couplings, etc.

This system ensures that the axis moves precisely and evenly. The electric motor provides the power to drive the motion, while the motion converter transforms this rotary motion into linear translation of the table or carriages. Limit stops and sensors are used to define the limits of the axis movements, ensuring the safety of the machining process. The encoder and tachometric dynamo are feedback elements that allow precise measurement of the position and speed of the axis in real time. Finally, the motion transmission elements ensure efficient, fluid transmission of the motor's motion to the table or carriages, ensuring stable, reliable axis movement.

## 1.5.9 Watering (or cooling) system

As with conventional machines, the coolant system is used in CNC machines to facilitate the cutting operation, cool the tool and workpiece, and improve the quality of machining. Most CNC machines are equipped with a high-pressure coolant system, generally between 70 and 80 bar, and up to 150 bar if conditions require.

This sprinkler system consists of a reservoir, a filter, a screen to prevent chips from entering the reservoir, one or more pumps, a distribution pipe and sprinkler nozzles.

In general, CNC machines are equipped with two coolant systems: an external system using hoses and an internal system. The internal coolant system is located at the centre of the spindle in milling machines and at the tool in turning machines.

The use of two coolant systems ensures efficient coolant delivery, tailored to the specific needs of each type of machine and machining process. This ensures optimum chip evacuation, adequate tool and workpiece cooling, and consequently improved overall machining quality in CNC machines.

## 1.5.10 Chip evacuation system

CNC machines are equipped with an automatic waste evacuation system, which prevents swarf from accumulating around the tool and workpiece. This system helps to reduce the time required to clean the machine manually.

CNC machine builders use various types of conveyor to evacuate waste. Among the most common are screw conveyors, chain conveyors, magnetic plate conveyors and scraper conveyors.

Each type of conveyor has its own advantages, depending on the specific machining requirements. Screw conveyors are valued for their ability to transport large quantities of swarf over long distances. Chain or magnetic plate conveyors are often used to evacuate metal chips, while scraper conveyors are effective for shorter, lighter chips.

Thanks to this automatic waste evacuation system, CNC machines can run continuously without interruption for cleaning, improving productivity and the overall efficiency of the machining process.

### 1.5.11 Tool change system

The production of mechanical parts in CNC machines requires the use of several cutting tools. To improve productivity and optimise the machining process, CNC machines are equipped with an automatic tool change system.

In the preparation phase, the tools required for machining are defined and prepared. They are then inserted into the machine's tool magazine. Machining centres are often characterised by the large number of positions available in their tool magazines. This extensive capability enables CNC machines to change tools quickly and automatically during the machining process, without the need for manual intervention.

This automatic tool change system offers a number of advantages, such as reduced downtime for tool change, improved production flexibility, and the ability to machine complex parts with a wide variety of tools, while maintaining high efficiency. Thanks to this functionality, CNC machines are able to carry out more sophisticated machining operations and optimise the production of mechanical parts with greater productivity.

### 1.5.12 Machining system

The machining system is responsible for performing the main function of the CNC machine. It consists of one or more spindles, a tool holder and a workpiece holder.

In CNC turning machines, the workpiece is clamped to the spindle by jaws. The rotation of the workpiece combined with the feed of the tool removes the material. Nowadays, there are CNC lathes with a single spindle, two spindles or several spindles, allowing simultaneous machining of a set of parts. The secondary spindle can be used to turn the part, making it easier to move on to the next phase. The part is then automatically placed in the secondary spindle (in the opposite direction) for further machining operations, optimising process efficiency.

In CNC milling, the workpiece is fixed on the table in the workpiece holder (vice, indexing table, special fixtures, etc.), while the tool is fixed in the spindle. The rotation and feed of the tool in conjunction with the movement of the workpiece removes the material.

CNC machining centres combine turning and milling processes, making it possible to manufacture complex parts in a variety of shapes.

The spindle is the key element of the machining system, as it generates the cutting movement. It is mounted on a pivot link in relation to the machine frame. To ensure machining accuracy, the spindle must be stable and dynamically balanced. In most cases, the pivot link is provided by ceramic ball bearings, although some machines use hydrostatic, hydrodynamic or magnetic bearings.

In CNC milling machines, the spindle rotation speed can reach 25,000 rpm. These milling machines are often equipped with high-frequency asynchronous AC motors, which are digitally controlled. On the other hand, low-speed spindles generally use DC motors. This diversity of configurations enables CNC machines to machine different materials and adapt to a wide range of machining requirements.

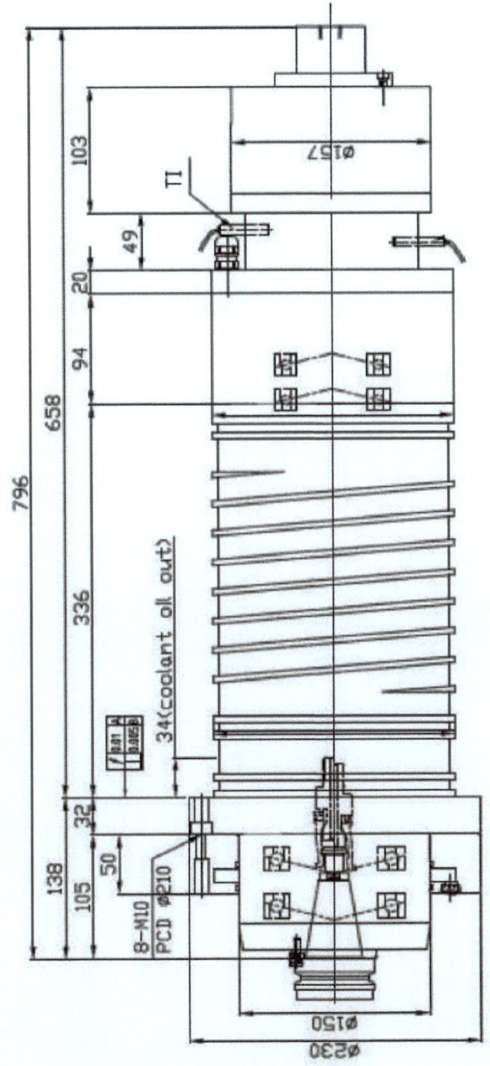


*Figure 6. Spindle for SPINNER machining centre[1]*

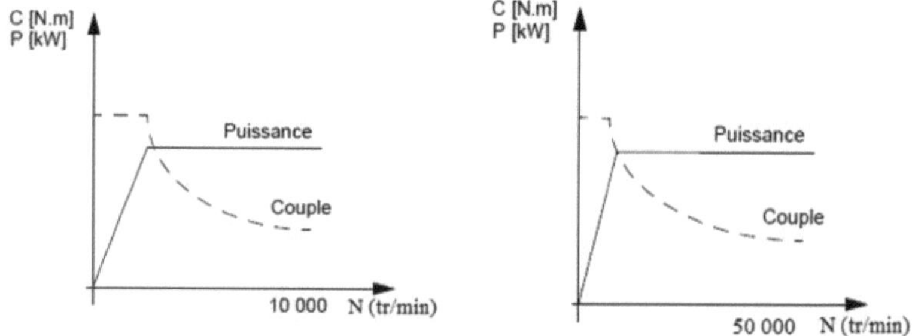


Spindle with a DC motor Spindle with a high-frequency asynchronous motor

## 1.6 CNC machine components
### 1.6.1 Components of a machining centre
The figures below show the main components of a CNC machine.
[1] Machine manual

[1] SPINNER machine manual

Electrical cabinet
Screen
Desk
Chip evacuation system
Door

Tool holder

7  :  Tool
8  :  Part
9  :  Vice (workpiece holder)
10  :  Table
11  Tool magazine
12  Tool change arm

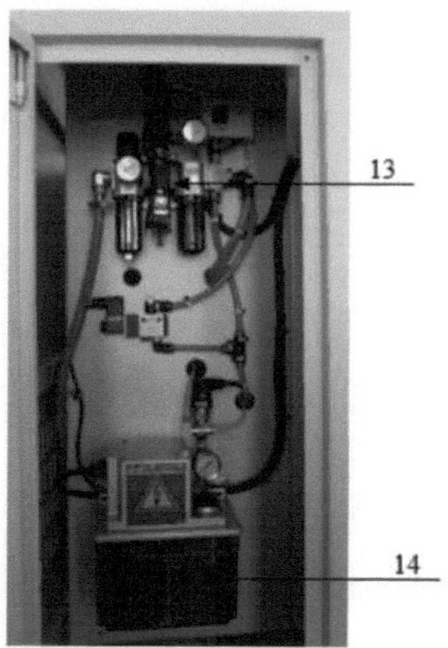

13 : Pneumatic circuit
14 : Hydraulic circuit

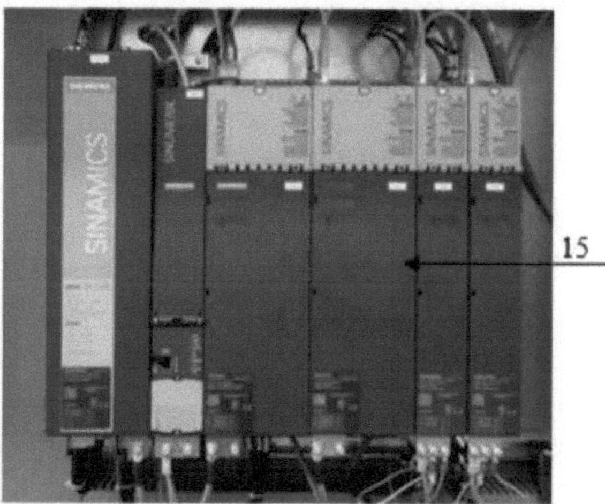

*Figure 7. Example of a SPINNER MVC 100 machining centre*

15  Programmable Logic Controller

**1.6.2 Components of a CNC lathe**

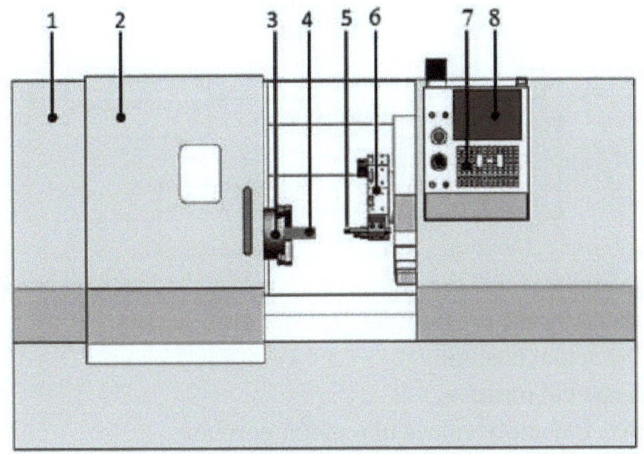

*Figure 8. Example of a CNC lathe*

1 : Bâti
2 : Door
3 : Spindle
4 : Part
5 : Tool
6 : Turret          tool    shop
7 : Pupitre
8 : Screen

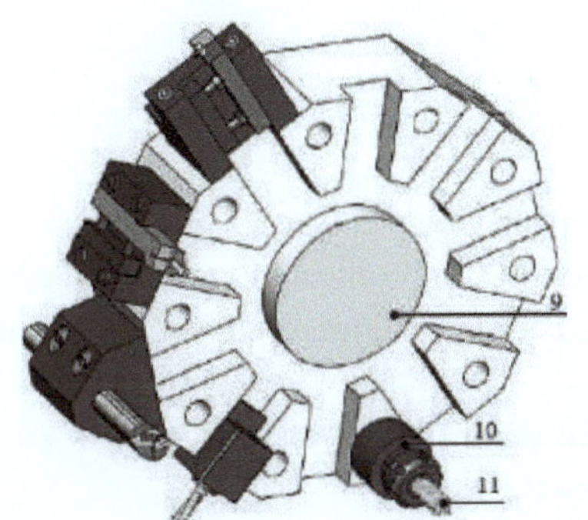

*Figure 9. Tool magazine on a CNC lathe*

9   : Turret
10  Tool holder
11  Tool

**1.7 MOCN servo-control**

**1.7.1 Position control**

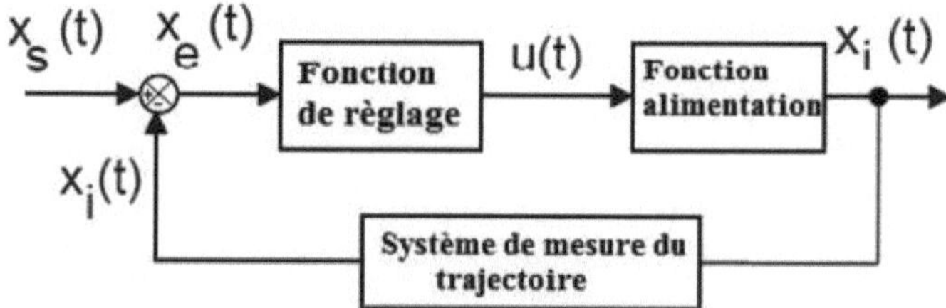


This system guarantees that the point controlled by the machine reaches the position requested by the programme. Its roles are :

- Monitor the actual position.
- Correct the actual position.
- Convert axis trajectory values into motor currents.

### 1.7.2 Open loop operation

In an open loop, the control unit moves the trolleys without any feedback on their actual position.

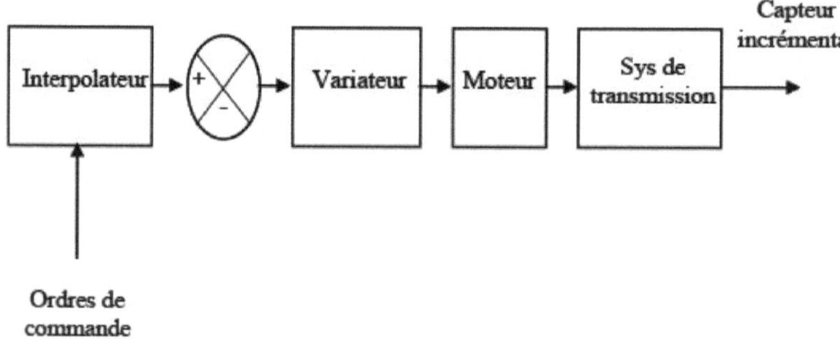


### 1.7.3 Closed loop operation

In a closed loop, the control unit controls the position and speed until the input setpoint (the command) and the measured setpoint (the actual position and speed) are equal.

The position of the point controlled by the machine is known by calculating the number of steps taken by the motor.

Displacement and rotation speeds are determined from the number of steps and the period. Depending on the position and speed required, the ECU increases or decreases the current sent to the motors.

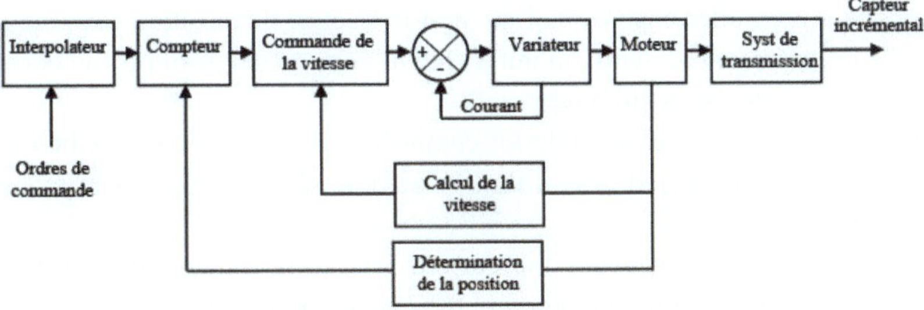


## 1.7.4 Operation with adaptive control

CNC machines are constantly evolving in order to achieve the best possible quality in the parts they manufacture. Nowadays, most of the parameters that influence machining are monitored. CNC machines are equipped with sensors that continuously measure torque, vibration, slippage, pressure, wear, temperature, and so on. The measured values are processed by an adaptive control director.

The machine control manager regularly adjusts the machine parameters to ensure the right working conditions for the operator, the machine and the tool. Thanks to this monitoring and constant adjustment of parameters, CNC machines are able to self-regulate and optimise machining performance, which in turn improves the quality of the parts produced and increases production efficiency. This also ensures greater tool and machine durability, as well as increased operator safety.

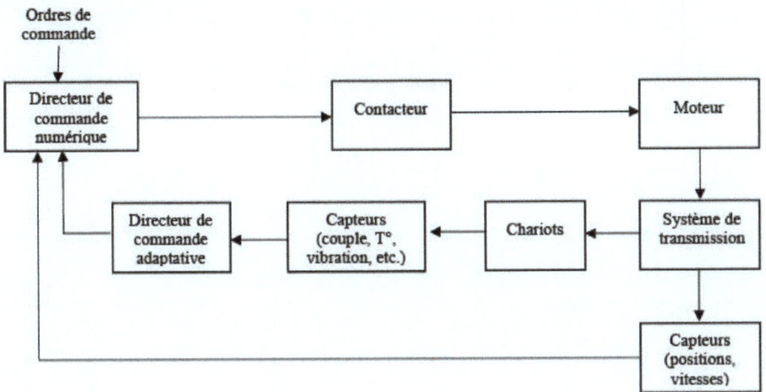


## 1.7.5 Motor control

Motor control in CNC machines is essential to ensure precise and efficient machining. To achieve this, the motors are controlled by various specialised controllers:

The position controller ensures that the motor reaches the position requested by the CNC program. It constantly monitors the motor's actual position and corrects any deviation from the position setpoint.

The speed controller is responsible for controlling the speed at which the motor moves. It adjusts the speed in real time to achieve the speed setpoint given by the CNC program.

The current controller regulates the current sent to the motor. It controls the force exerted by the motor and therefore the machining power.

Actual measurements of position, speed and current are obtained using sensors placed on the motor axes. These measurements are then used by the controllers to make the necessary corrections and ensure optimum machine operation.

The setpoints, also known as final interpolation, are the instructions given to the controller to define the axis paths. These setpoints determine the overall movement of the machine and are based on the commands in the CNC program. Thanks to these setpoints, the machine can carry out complex machining operations with great precision and efficiency.

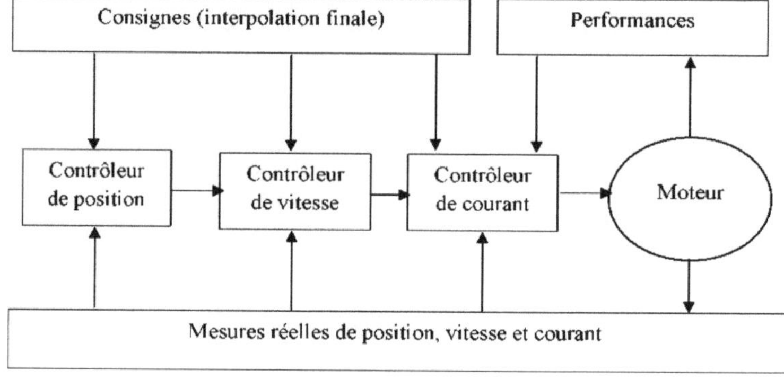

*Table 1. Control of DC motors*

| **Voltage controlled** | **Current controlled** |
|---|---|
| **Commandé en tension** | **Commandé en courant** |
|  |  |
| $U_A(t) = R_A \cdot i(t) + L_A \dfrac{di(t)}{dt} + U_E(t)$ (Eq.1) | $U_A(t) = R_A \cdot i(t) + L_A \dfrac{di(t)}{dt} + U_E(t)$ (Eq.4) |
| Couple d'entraînement | Couple d'entraînement |
| $M(t) = K_M \cdot i(t)$ (Eq.2) | $M(t) = K_M \cdot i(t)$ (Eq.5) |
| Couple d'accélération | Couple d'accélération |
| $J \dfrac{d\omega(t)}{dt} = M(t) - M_L(t)$ (Eq.3) | $J \dfrac{d\omega(t)}{dt} = M(t) - C_R\,\omega(t)$ (Eq.6) |

### Control error

Control errors can occur for a variety of reasons, such as electromagnetic interference, hardware faults, communication problems between system components, or controller programming errors.

When a control error occurs, it can lead to defects in the machined parts, deviations from the required dimensions and specifications, and even machine stoppages to prevent further damage.

To minimise motor control errors, CNC machine manufacturers are implementing monitoring and redundancy mechanisms, advanced diagnostic systems and safety devices to stop the machine if anomalies are detected.

CNC machine operators must also be trained to detect and react to possible control errors, by carrying out regular checks, monitoring indicators and applying the appropriate procedures in the event of a problem.

Despite the precautions taken, it is important to understand that control errors can occur and that it is crucial to implement measures to prevent and manage them effectively to ensure high quality machining and smooth production.

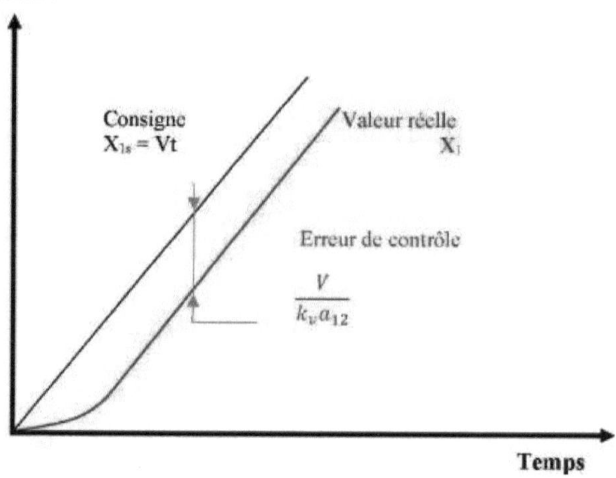

<u>**DC motor control loop (current-controlled) :**</u>

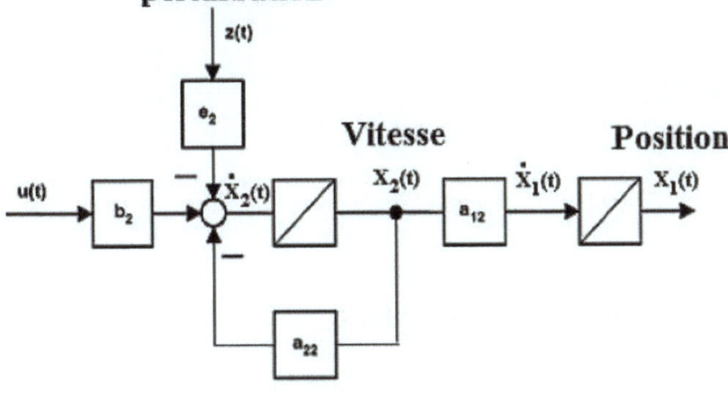



$$\dot{X}(t) = A \cdot X(t) + b \cdot u(t) + e \cdot z(t)$$

(Eq.7)

Avec :

$$a_{12} = \frac{\omega_0}{\varphi_0}$$

$$a_{22} = \frac{C_R}{J}$$

$$b_2 = \frac{k_m \cdot i_0}{\omega_0 \cdot J}$$

$$e_2 = \frac{m_0}{\omega_0 \cdot J}$$

<u>**DC motor control loop with speed controller :**</u>

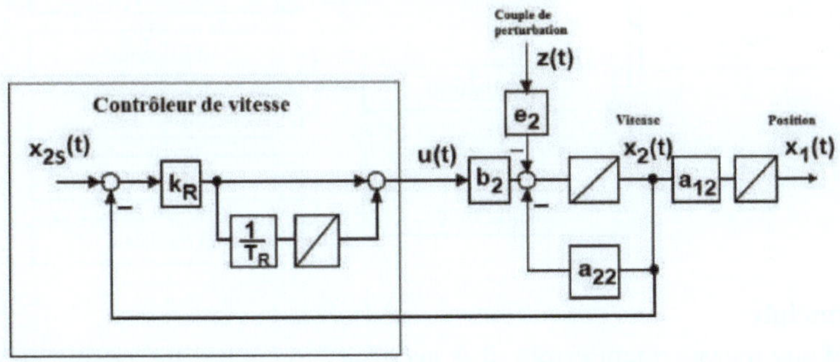



<u>**Control loop for a DC motor with a
and a position controller :**</u>

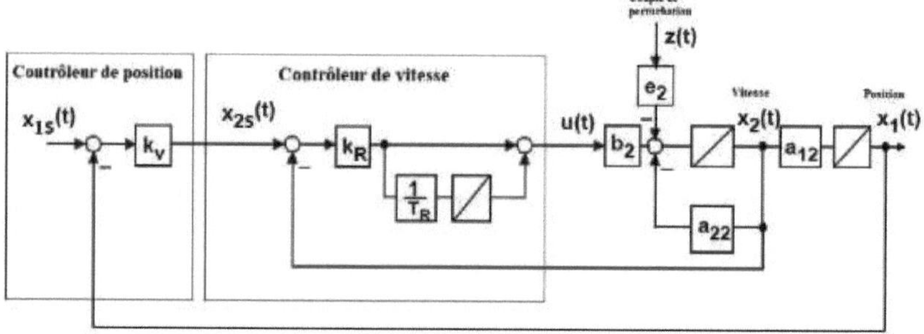

### Digital position and speed controller :

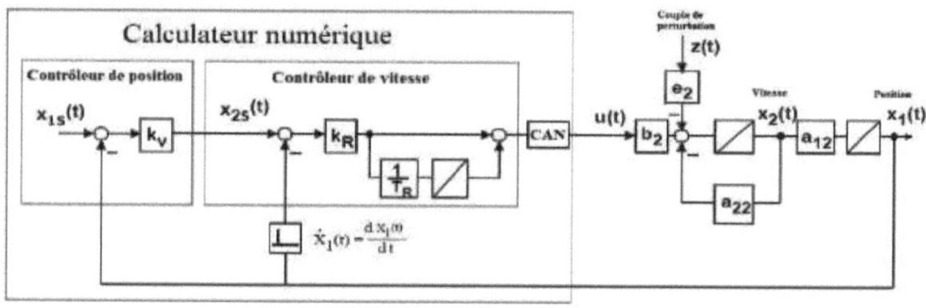

## 1.8 Measuring trajectories

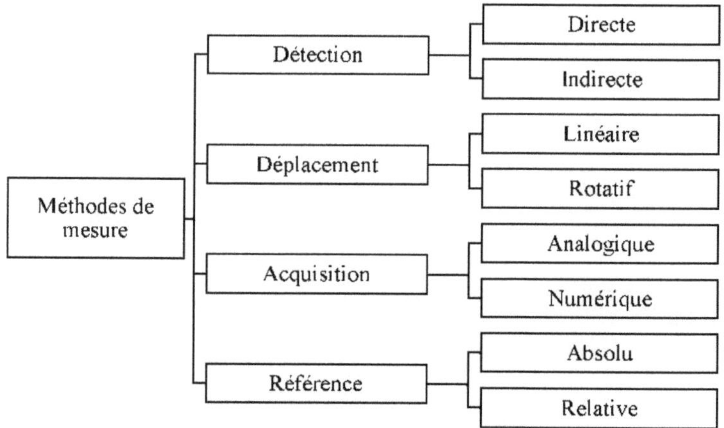


### 1.8.1 Principle

To avoid any measurement errors, it is necessary to control the positions of the carriages in real time. Each axis (linear or rotary) is fitted with one or more sensors called encoders. The encoders play a crucial role in determining the precise position of the carriages.

An encoder generally consists of a transistor biased by an LED that emits light. This light is directed towards another transistor. If the second transistor receives light, it closes, acting like a closed switch. On the other hand, if it does not receive light, it remains blocked, acting as an open switch.

The control part of the CNC machine determines the position by counting the number of signals received by the encoder. By measuring the number of signals sent and received by the encoder, the controller can calculate the current position of the axis with a high degree of accuracy.

By monitoring the positions of the carriages in real time using encoders, the CNC machine can correct any deviation from the programmed path, ensuring precise machining to the required specification. This approach avoids measurement errors and guarantees optimum quality in the production of mechanical parts.

*Figure 10. Encoder*

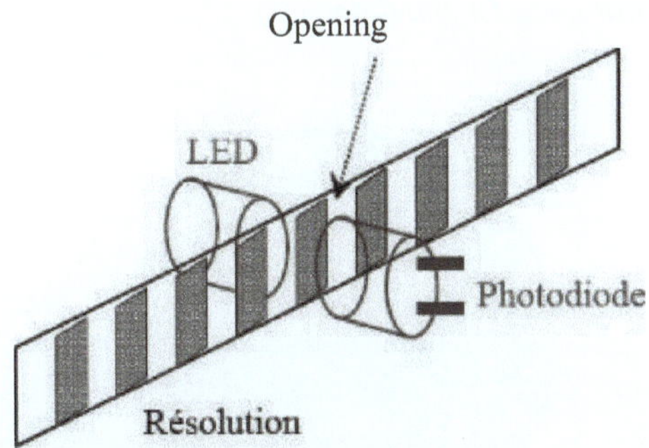


*Figure 11. Operating principle of an encoder*

**1.8.2 Incremental measurement**

A (relative) incremental encoder sends out a set of pulses for each revolution of the motor, and the number of pulses is then converted into a displacement value (linear or angular). This measured value is then added to the previous value to track the movement of the axis.

Numerically controlled machine tools fitted with these encoders require an initial reference setting procedure after each power-up. This procedure defines a reference position from which movements are measured and controlled.

Most of these encoders have two rows of apertures, called "A" and "B", which are 90 degrees out of phase. This phase shift makes it easier to determine the direction of movement of the shaft. When the axis moves in one direction, the "A" and "B" pulses are offset, allowing the controller to determine the direction of movement.

The encoder also includes a segment marked "Z", which is used to determine the number of complete revolutions made by the shaft. This segment provides a specific pulse for each complete revolution of the motor, enabling the controller to track the total number of revolutions made.

Thanks to the information provided by the encoder, the CNC machine controller can monitor the position of the axis in real time and make the necessary corrections to precisely achieve the programmed path, ensuring reliable, high-quality machining.

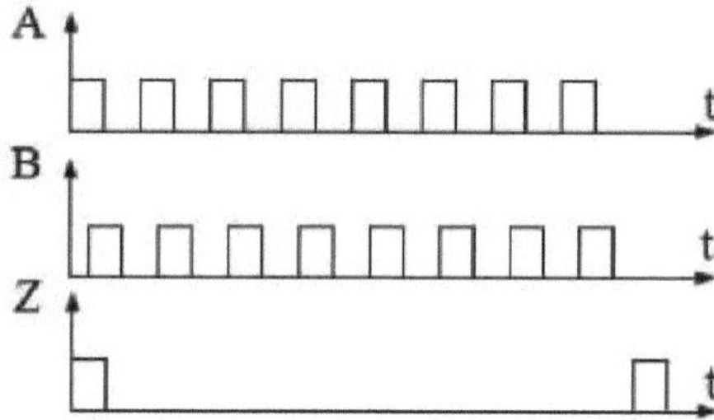

*Figure 12. Incremental encoder timeline*

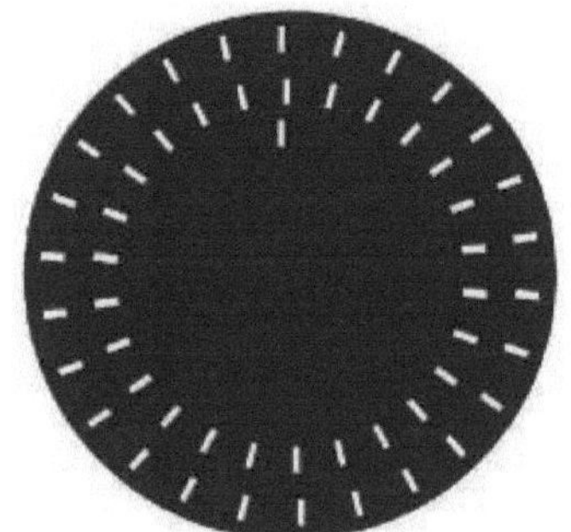

*Figure 13. Incremental encoder*

### 1.8.3 Absolute measurement

Nowadays, numerically controlled machines (NCMs) are increasingly equipped with absolute encoders. An absolute encoder can determine the exact coordinates of the point controlled by the machine at any time, even after power-up. NC Machines using this type of encoder do not require a referencing procedure after power-up, as the absolute encoder immediately provides the current position of the axis.

The absolute encoder uses a disc with rows of apertures (n bits). Each row has its own distribution of apertures, and at each position of the disc, it gives a specific binary code representing the precise position of the axis.

These absolute encoders require a high number of bits to encode all possible axis positions, as each row of apertures corresponds to a specific part of the position. This means that they can be more expensive and require more energy to encode and process the data.

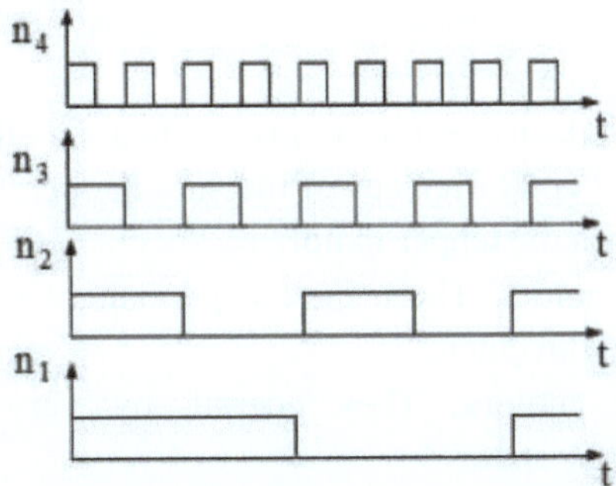

*Figure 14. Chronogram of an absolute encoder*

However, despite their higher cost and power consumption, absolute encoders offer the advantage of providing accurate and instantaneous measurement of axis position, eliminating the need for a referencing procedure and making it easier to resume work after a stoppage or power cut. These features make absolute encoders the preferred choice for many applications where precision and reliability are essential in machining and industrial production.

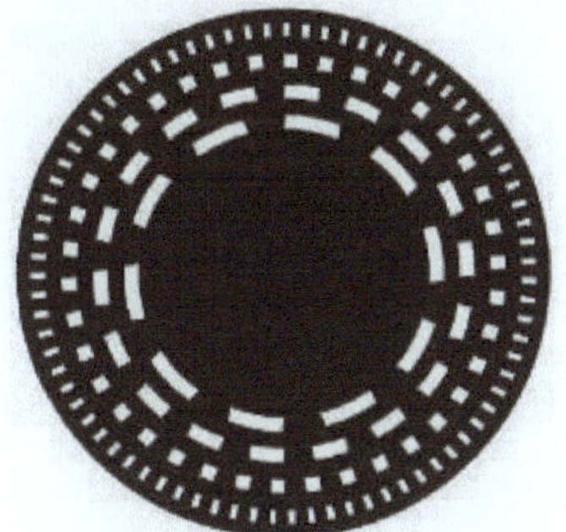

*Figure 15. Absolute encoder*

## 1.9 The engines

### 1.9.1 Stepper motors

Stepper motors are devices that transform electrical control pulses into rotation, taking a specific number of steps with each signal. They are characterised by their resolution, i.e. the number of steps they take per complete revolution. This resolution has a direct impact on the precision of the CNC machine and, consequently, on the quality of the parts manufactured.

However, stepper motors have a number of disadvantages:

- They are relatively inefficient compared with other types of motor.
- To obtain smooth movements, they require the use of the 'microstepping' technique, which consists of making finer subdivisions of the steps. This improves the fluidity of the movements.
- Their acceleration capacity is not as fast as that of other types of motor, which can limit the speed and efficiency of machining operations.

There are three main types of stepper motor:

- Permanent magnet motors: These have a permanent magnet to create the magnetic field and drive movement.
- Variable reluctance motors: They operate by varying the reluctance (magnetic resistance) in the stator.
- Hybrid motors: These combine the characteristics of permanent magnet motors and variable reluctance motors, offering improved performance.

Despite their disadvantages, stepper motors are still widely used in CNC machines due to their relatively low cost, simplicity and sufficient accuracy for many applications. However, in some applications requiring high speed and power, other types of motor, such as DC or AC motors, may be preferred.

*Figure 16. Stepper motor for CNC machine*

### 1.9.2 Servomotors

AC servomotors are becoming increasingly popular in CNC machines. These motors operate in a closed loop, which means that their position is continuously checked and corrected according to measured values. This feedback ensures high precision and reliability in machining.

Servomotors used in CNC machines are generally controlled by a set of commands that flow through a digital bus. Unlike the analogue pulses used in some other types of motor, servo motors use digital servo control. This allows more precise and responsive control of the motor's position and speed, resulting in better machining quality.

Servomotors are renowned for their high efficiency and rapid acceleration. They are able to start and stop quickly, which reduces cycle times and improves the productivity of the CNC machine.

Thanks to their continuous feedback and digital servo control, servomotors offer a high-performance solution for applications requiring high precision and rapid response to changing cutting conditions. Their use helps to improve the quality of the parts produced and increase the overall efficiency of CNC machines.

The disadvantages are :

- More expensive than stepper motors.
- More complicated to use.
- They degrade rapidly as a result of overheating and overloading.
- They require more maintenance.

They are made up of :

- A DC motor.
- A speed reducer.
- A potentiometer to generate a variable current.
- A servo-control device.

*Figure 17. Servomotor for NC machine*

## 1.10  Variable speed drives

Variable speed drives are electronic devices used to regulate the speed of electric motors. In most numerically controlled (CNC) machines, variable speed drives can be controlled in two ways: either manually using potentiometers (one for slide feed speed and another for spindle speed), or by the machining programme itself.

To adjust the speed of an electric motor, variable speed drives act on two parameters: the armature supply voltage or the flux produced by the inductors.

When the voltage is varied with a constant flux, the motor torque remains at a constant value while the power generated can be increased or reduced. The speed of rotation then varies in proportion to the power.

On the other hand, when the flux is modified with a constant voltage, this leads to a variation in motor torque.

New generations of CNC machines use synchronous and asynchronous AC motors, combined with frequency inverters that use pulse-width modulation (PWM) and insulated-gate bipolar transistors. These technologies enable more precise control of motor speed and power, helping to improve the overall performance of CNC machines. By using variable frequency drives, CNC machines can adapt efficiently to different machining needs and optimise productivity while maintaining high precision in machining operations.

*Figure 18. Variable speed drive*

**Features** :

> Rotating field speed :

$$n_s = \frac{s}{p} \qquad (Eq.8)$$

**ns** : Rotating field speed (synchronism) in rpm.

**f**: Frequency in Hz.

**P**: Number of pole pairs, without unit.

> Rotation frequency

$$n = n_s\,(1 - g) \qquad (Eq.9)$$

Where **g** is the slip

$$n = \frac{(n_s - n)}{n_s} \qquad \text{(Eq.10)}$$

$$n = \frac{f}{p(1-g)} \qquad \text{(Eq.11)}$$

The technological solution for controlling the speed of motors in CNC machines is to use frequency converters. These devices enable the frequency of the electric current supplied to the motor to be varied, which directly influences its speed of rotation.

Frequency converters allow the frequency of the current supplied to the motor to be adjusted within a range from 0 Hz (complete standstill) to the frequency corresponding to the motor's rated speed (for example, 50 Hz for a motor used in a region where the frequency of the electric current is 50 Hz).

This ability to vary the frequency of the current allows precise control of the motor speed, offering great flexibility in machining operations. CNC machines can be configured to operate at different speeds according to the specific needs of each machining process.

In addition, certain types of motor, particularly asynchronous motors, can be operated at speeds higher than their rated speed thanks to this frequency converter technology. This makes it possible to achieve overspeeds that can be useful in certain machining applications.

By using frequency converters, CNC machines benefit from greater adaptability and precision in controlling motor speed, which helps to improve the efficiency and quality of the machining carried out.

> Schematic diagram

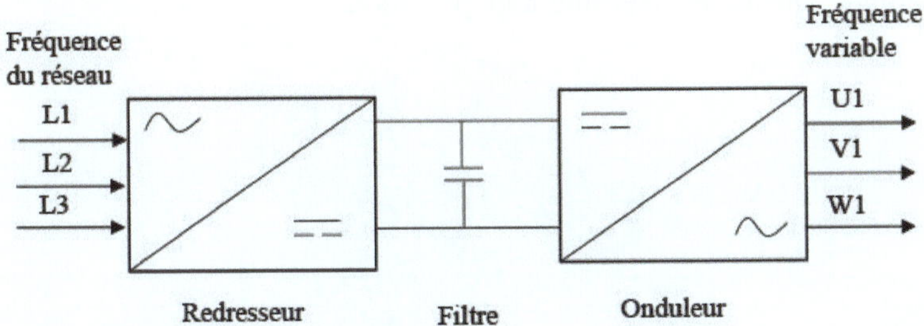


## 1.11 Motion transmission systems

The most commonly used systems for converting rotational movement into linear movement of carriages are :

- Ball screw and nut system.

Rack and pinion.

Belt pulleys.

Linear motor.

The choice of conversion system depends on the specific needs of each application and the requirements in terms of precision, load, durability and cost. Each system has its advantages and disadvantages, and it is important to select the one that best meets the constraints and desired performance for the machining operations carried out by the CNC machine. The table below compares the systems for converting rotary motion into linear movement of carriages:

*Table 2. Motion transmission systems*

|  | **Rack and pinion** | **Ball screw nut** | **Linear motor** |
|---|---|---|---|
| **Accuracy(mm)** | 0,01 | 0,005 | 0,005 |
| **Maximum length of linear travel (m)** | Unlimited | 10 | Unlimited |
| **Cost** | Low | High | Very high |
| **Preferred use** | CNC machines with long linear movements and high loads. | CNC machines with shorter linear travels and medium loads. | CNC machines requiring high precision linear movements. |

### 1.11.1 Ball screw and nut system

The 'ball screw' conversion system is widely used in CNC machines because of its advantages in terms of precision, speed and load capacity. This helical linkage mechanism, with balls interposed between the screw and the nut, significantly reduces friction compared with a conventional screw-nut system. Reduced friction improves the accuracy of linear movements, which is essential for machining operations requiring high precision.

*Figure 19. Ball screw nut*

Numerically controlled machine tools using the ballscrew system offer advantages such as faster, smoother linear movements, greater repeatability of movements, greater system rigidity and lower component wear. These features make them the preferred choice for applications requiring high precision, high speeds and heavy loads.

Because of its improved performance, the ballscrew system is often preferred in CNC machines to guarantee optimum manufacturing quality and increased efficiency in machining processes.

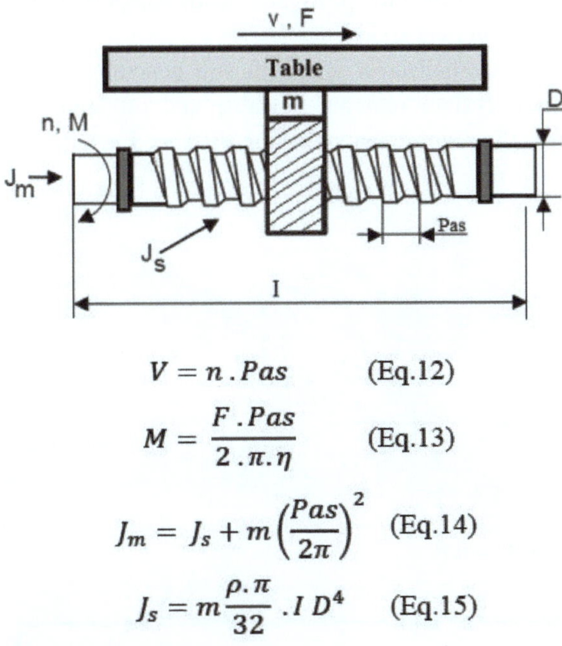


$$V = n \cdot Pas \qquad \text{(Eq.12)}$$

$$M = \frac{F \cdot Pas}{2 \cdot \pi \cdot \eta} \qquad \text{(Eq.13)}$$

$$J_m = J_s + m \left(\frac{Pas}{2\pi}\right)^2 \qquad \text{(Eq.14)}$$

$$J_s = m \frac{\rho \cdot \pi}{32} \cdot I \, D^4 \qquad \text{(Eq.15)}$$

**V:** Table (carriage) linear travel speed, in m/s.

**n:** Motor speed, in rpm.

**M:** Motor torque, in Nm.

**m:** Mass of the table, in Kg.

Js: Moment of inertia of the screw, in Kg.m2.

Jm: Total moment of inertia, in Kg.m2.

*p*: Density (steel p = 7850 kg / m ).[3]

**D:** Screw diameter, in metres.

**I:** screw length, in metres.

### 1.11.2 Rack and pinion system

This 'rack and pinion' system is another conversion mechanism used in some CNC machines. It consists of a toothed wheel (pinion) and a rod with teeth of the same module (rack). These two elements mesh so that rotation of the pinion causes linear movement of the rack without slippage.

This system is mainly used in CNC machines requiring very long linear travel distances. The rack acts as a 'ramp' along which the pinion moves, transforming the rotational movement of the pinion into a continuous linear movement of the rack.

Although the rack and pinion system is effective for long linear movements, it is less commonly used than the ballscrew system, as it can be less precise and has certain limitations in terms of speed and loads supported. For applications requiring high precision over short

At longer distances, the ballscrew system is still generally preferred. However, the rack-and-pinion system remains a viable and effective option for CNC machines with specific requirements for long-distance linear motion.

*Figure 20. Rack and pinion*

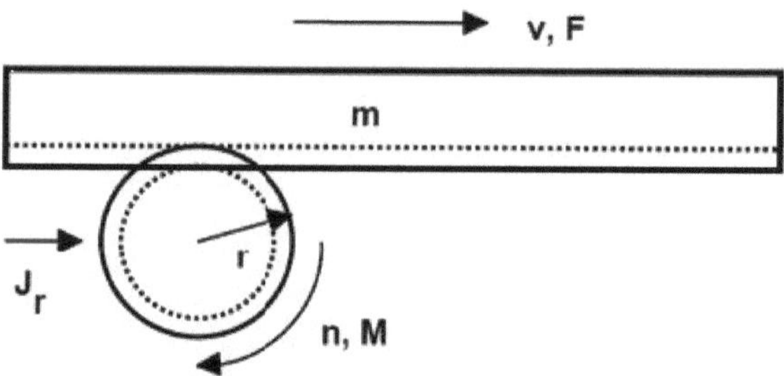


$$V = n\,.\,2\,.\,\pi\,.\,r \qquad (\text{Eq.16})$$

$$M = \frac{F\,.\,r}{\eta} \qquad (\text{Eq.17})$$

$$J_m = J_r + m.r^2 \qquad (\text{Eq.18})$$

$$J_s = \frac{\rho.\pi}{32}\,.\,I\,D^4 \qquad (\text{Eq.19})$$

**V:** Table (carriage) linear travel speed, in m/s. **n:** Motor rotation speed, in rpm.

**r**: Radius of gable, in metres.

**M**: Motor torque, in Nm.

**m**: Mass of the table, in Kg.

**Jr**: Moment of inertia of the gable, in Kg.m2.

Jm: Total moment of inertia, in Moment Kg.m2. *p*: Density (steel $p = 7850$ kg /
m3). **D**: Diameter of pinion, in metres.

**I**: length of gable, in metres.

### 1.11.3 Linear motor

Linear motors are a variation of servomotors in which the rotor and stator are arranged flat, enabling them to directly generate a linear translational force rather than a rotary motion. Unlike other motion conversion systems, linear motors do not require an additional mechanism to transform rotational motion into linear displacement. Linear motion is generated by the electromagnetic interaction between the moving part (which consists of coils) and the fixed part (which consists of permanent magnets).

In CNC machines, the moving part of the linear motor is attached directly to the machine carriages, eliminating the mechanical play often associated with traditional motion transmission systems. This design allows greater precision in the positioning and movement of the carriages, which is essential for the quality and precision of the machining performed by the CNC machine.

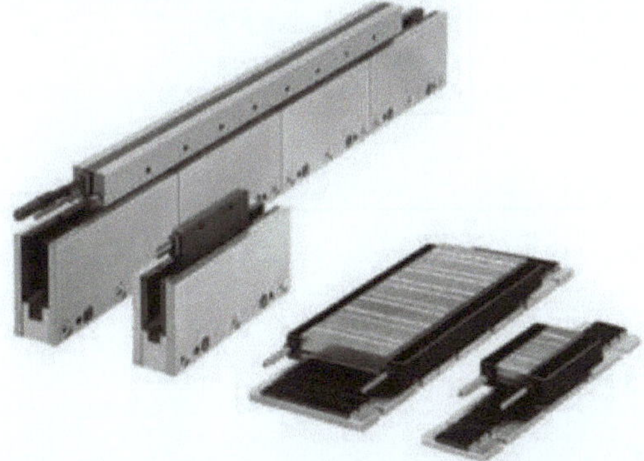

*Figure 21. Linear motors for HEIDENHAIN CNC machines*

Linear motors are particularly popular in CNC machines for their ability to provide fast, accurate linear movement with no mechanical backlash, making them suitable for applications where high precision and high dynamics are required .

### 1.12 MOCN programming methods

There are several ways of controlling CNC machines:

- **Manual programming :**      This method consists of  writing manually the movement instructions needed to machine the workpieces. These instructions are entered via the machine's control panel or using a keyboard. Each move is specified point by point, meaning that the programmer must enter the precise co-ordinates of the end position for each machine movement. Each move is recorded in a separate block (or line) of the program. The programmer writes a sequence of blocks that define all the movements required to machine the part. Once the programme is ready, the operator presses the cycle start button to start the CNC machine executing the programme. The CNC machine then executes the programme blocks in chronological order, following the specified movement instructions. It moves from point to point following the co-ordinates entered by the programmer, making the movements required to machine the part according to the programme specifications. This method is widely used in workshops that do not need to produce large runs of identical parts, or that require fine, precise adjustments for each machining operation.

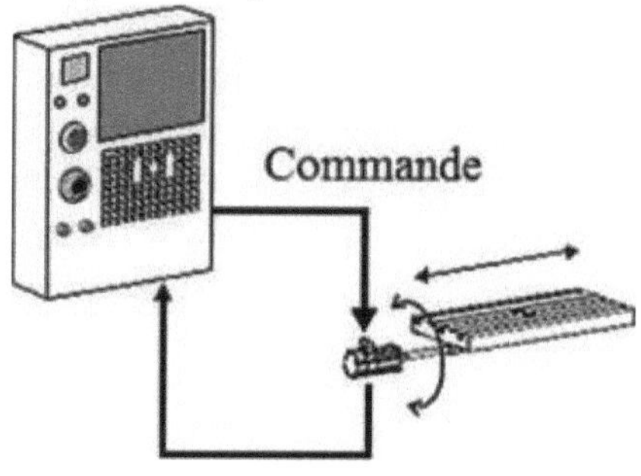


```
N10 G90 G40 G71
N20 T1D1 M6
N30 G97 S800
N40 G94 F200
N50 G00 X-15 Y10 Z5 M3 M7
N60 G01 X0 Z-3
N70 G02 X10 Y10 R5
N80 G00 G52 X0 Y0 Z0 M5 M9
N90 M2
```

*Figure 22. Manual programming*

**- Geometric Profile Programming (PGP):** This method of machining complex CNC profiles, also known as manual programming using simple geometric elements, is used to machine parts with complex paths. This method consists of breaking down the desired trajectory into simple geometric elements, such as straight lines, circular arcs or reference points, and then specifying the relationships and parameters between these elements.

The CNC programmer defines each geometrical segment by specifying the start and end coordinates, the radii of the circular arcs, the angles of inclination, and so on. By linking these simple geometric elements together, we obtain the complete tool path for machining the part to the desired profile. This method is often used to machine complex contours, irregular shapes or parts with particular geometries. It offers great flexibility and precise control over machining paths.

However, it can be more laborious and require some CNC programming expertise to break down the path properly. The advantage of this method is that it allows the programmer to have direct control over each element of the path, which can be useful for achieving precise finishes or for performing specific operations that are not easily achievable with other CNC programming methods.

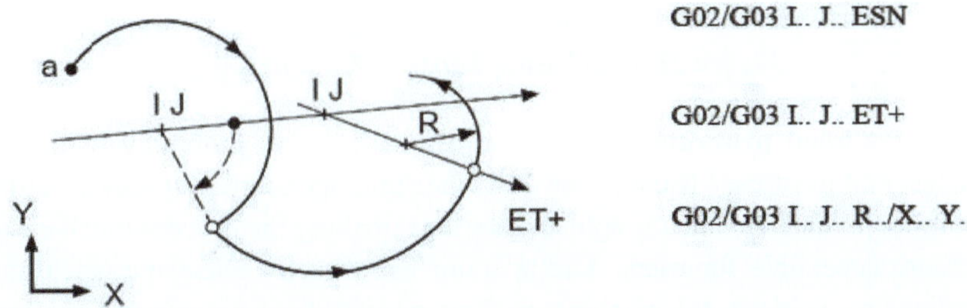


- **Computer Aided Programming (also known as CAM):** To facilitate CNC programming and make the process more efficient, programmers generally use Computer Aided Manufacturing (CAM) software. This software automatically generates the machining programme from a 2D or 3D CAD (Computer Aided Design) model. The programmers define the machine parameters, the machining operations required and the tools to be used in the CAM software. The software then analyses the CAD model and automatically generates the machining paths required to produce the desired part. This saves programmers the tedious task of manually determining the point coordinates for complex paths.

As well as generating the machining programme, CAM software also allows machining operations to be simulated to detect collisions, path errors or any other potential problems before actual production begins. This prevents costly accidents and improves the safety of CNC machines. Once the machining program has been created and simulated in the CAM software, it can be transferred to the CNC machine either via a connection cable, USB stick or internet connection, depending on the machine configuration.

There are many CAM software packages on the market, each with its own features and functionality. Some of the popular CAM software includes TOPSOLID, MASTERCAM, CATIA, SOLIDCAM, ONE CNC, AUTODESK

FEATURECAM, XCAP, and many others. Companies generally choose the CAM software that best suits their specific needs based on the type of machining they do and the complexity of their parts.

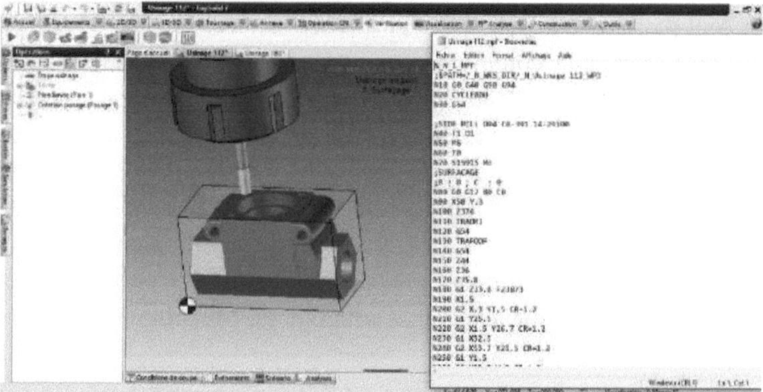

*Figure 23. CAM using TopSolid7® software*

**- Conversational programming:** The conversational programming method is the latest and most user-friendly for CNC machine operators. It involves setting machining parameters via a graphical interface, making the process much easier and more accessible for users. The operator can directly select the machining operations available in the machine, such as turning, facing, surfacing, drilling, tapping, pocket machining, path machining, grooving, etc., or use standard functions such as repeat, mirror, tool change, etc.

In this method, the operator does not need to write any programming code, as he simply enters the values of the specific cutting parameters for each operation. For example, for a drilling operation, the operator enters the speeds, position and depth of the hole, and the planes of attack and clearance, and the machine takes care of converting these parameters into machining movements. Using conversational programming does not require in-depth knowledge of CNC

The use of conversational programming does not require in-depth knowledge of CNC programming, as the machine's control unit takes care of converting the parameters into axis movements. This makes the machining process faster and more accessible for less experienced operators.

It is important to note that conversational programming is often limited to simple, standard operations. For more complex operations or customised trajectories, it may be necessary to use more advanced and comprehensive programming methods.

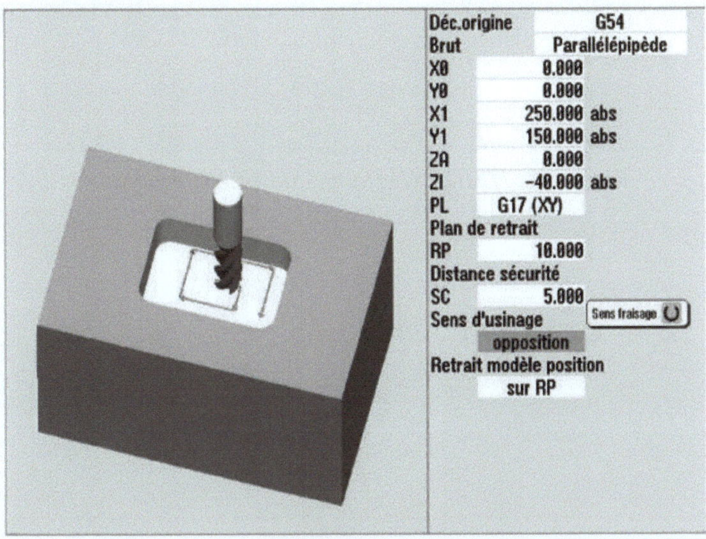


*Figure 24. SHOPMILL® bag settings screen*



*Figure 25. SHOPMILL SPINNER® programme*

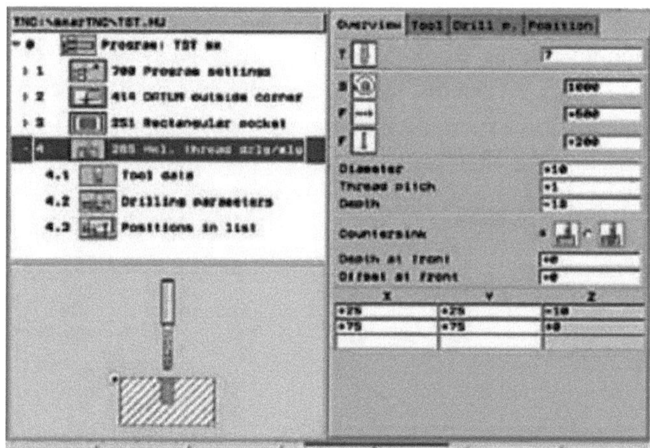

*Figure 26. HEIDENHAIN® SMART.NC program*

# Chapter 2

# Use of numerically controlled machine tools

## 2.1 Introduction

Before moving on to programming, it is essential to define the reference marks and datums that will serve as the basis for machining operations. These reference points define the positions of the part and the tool, as well as the axes of movement.

The origin point is a fundamental reference point for all machining operations. It is usually located at a corner of the workpiece or at a specific point that facilitates calculations and movements. The coordinates of the other points on the workpiece and the tool movements are determined from this origin point.

Axes are also important markers that define the directions in which the machine moves. In CNC machines, the X, Y and Z axes are generally used to define movements along the horizontal and vertical axes. In some cases, machines with more axes can be used, allowing more complex movements and more sophisticated operations.

The choice of reference marks and axes depends on both the machine equipment and the nature of the operations to be carried out. It is important to define these reference frames correctly to ensure the accuracy and consistency of the machining operations. Once the reference frames have been defined, the programmer can begin to write the CNC program using the appropriate coordinates and movements in the chosen reference frame.

## 2.2 Reference points

### 2.2.1 Origins

**OM** "Machine Origin": this origin is defined by the positions of the electrical stops in the machines.

**Om** "Measurement origin": this origin is specified by the CNC machine manufacturer and represents the first zero point of the encoder; it is an origin relative to the machine origin. If necessary, it can be reset by the operator according to a procedure set by the manufacturer. In most CNC machines, manufacturers set the measurement zero point at the same position as the machine zero point.

**Opp** "Workpiece holder origin": this is the link between the machine and the workpiece holder, in relation to the measurement origin. It represents the relative position of the workpiece holder.

**Opo** "Tool holder origin": this is the point controlled by the machine without tool correction. It is defined by the machine manufacturer in relation to the measurement origin.

Workpiece origin": this is the link between the workpiece holder and the workpiece, and specifies the position of the workpiece in the workpiece holder. This origin is chosen by the operator according to the shape and dimensions of the part.

**OP** "Program Origin": this is the origin of all programmed tool and/or part movements. It is defined in relation to the other origins. Its position is freely chosen by the programmer, based on the part's dimensioning system. Obviously, it is possible to position the program origin at the same position as the part origin.

*Table 3: Origins*

| Origins | Symbols |
|---|---|
| Machine origin | |
| Measurement origin | |
| Part origin | |
| Programme origin | |

### 2.2.2 Machine Piloted Point (MPP)

This is the starting point for tool measurement, defined by the machine manufacturer and located on the tool holder system. The operator can move this point in relation to the measurement and programme origins.

### 2.2.3 Cutting edge (AT)

The cutting edge is where the cutting operation takes place. It is defined in relation to the point controlled by the machine (PPM) and depends mainly on the geometry and dimensions of the tool. Before machining, the operator must define each tool by its designation and its own offset. The offset sets the distances between the cutting edge and the PPM in relation to each axis.

In most cutting tools, the cutting edge is considered to be a reference point located at the end of the tool. This point will become the point controlled by the machine controller. During the machining operation, whether it is the tool or the workpiece that moves, the controller will always act as if it were moving the cutting edge relative to the program origin. So you need to program in relation to this point, because what matters is not how the workpieces move, but how the cutting edge moves in relation to the work.

**AT symbols :**

For turning: For milling :

## 2.3 The axes

In a numerically controlled machine tool, axes refer to the linear and rotary movements of a machine component, such as the table, slide and spindle. The axes in a CNC machine are defined in accordance with ISO 841 and AFNOR NF Z 68-020. Axis markers are always located on the machine tools:
- On the spikes for the towers.
- To the strawberry centres.

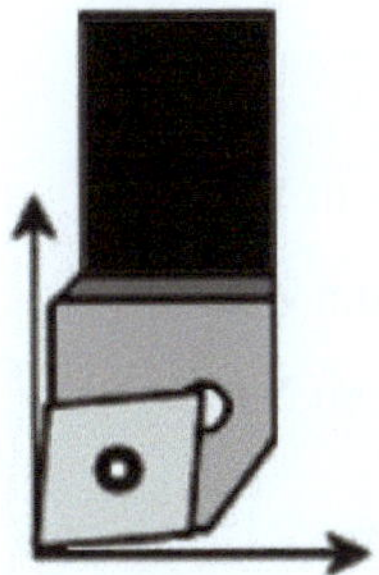

Figure 27. Position of the "turning tool" axes

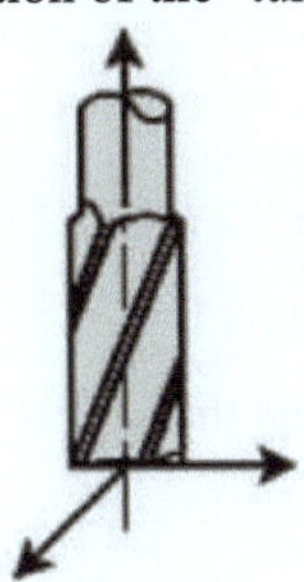

*Figure 28. Position of "milling tool" axes*

**Primary axes :**

❖ The **Z** axis corresponds to the spindle axis, with a positive direction

45

corresponding to an increase in the distance between the workpiece and the tool.

❖ The **X** axis corresponds to the next axis with the greatest displacement, the positive direction corresponds to an increase in the distance between the workpiece and the tool.

❖ The **Y** axis forms, with the other two axes "X and Z", a trirectangular trihedron with a direct direction "the rule of the three fingers of the right hand".

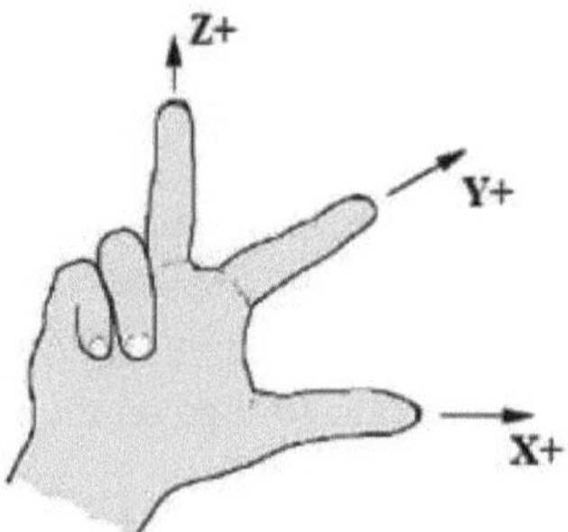


*Figure 29. Right-hand ruler*

## The axes of rotation :

❖ **A** around **X**, direction **A+** from **Y** to **Z**

❖ **B** around **Y**, direction **B+** from **Z** to **X**

❖ **C** around **Z**, direction **C+** from **X** to **Y**

## Secondary routes

The secondary translation axes are :

❖ **U** parallel to the **X** axis

❖ **V** parallel to the **Y** axis

❖ **W** parallel to the **Z** axis

The secondary axes of rotation are :

❖ **D** coaxial with axis **A** around axis **U**

❖ **E** coaxial with the **B** axis around the **V** axis

## Tertiary routes

The tertiary translation routes are :

❖ **P** parallel to the **X** axis

❖ **Q** parallel to the **Y** axis

❖ **R** parallel to the **Z** axis

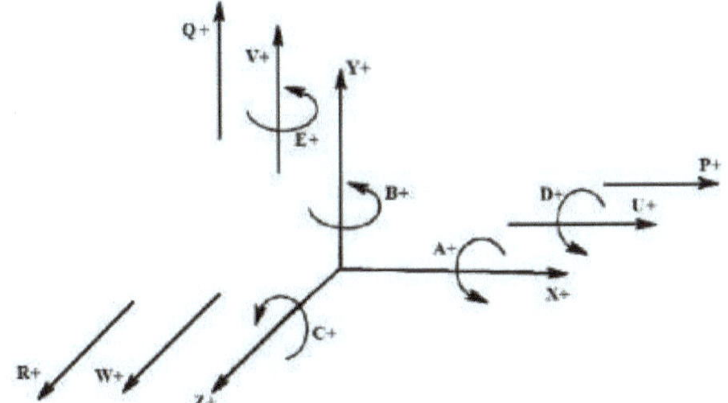

*Figure 30. Axis system on CNC machines*

**Note:** in cases where the displacements are given in relation to the workpiece, the directions of the axes will be reversed and the designation will be noted: X', Y', Z', A', ...

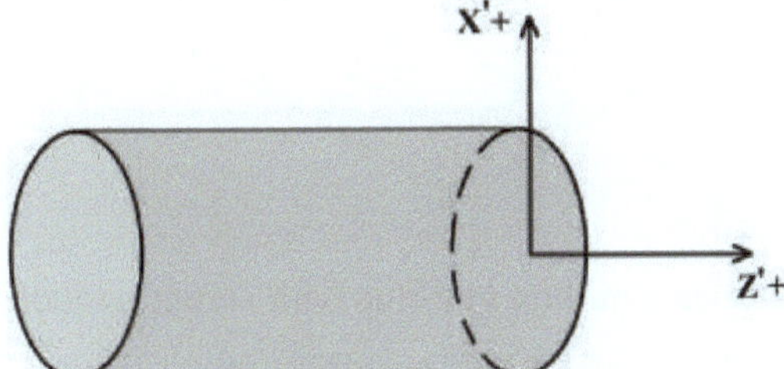

*Figure 31: Axes on a cylindrical part*

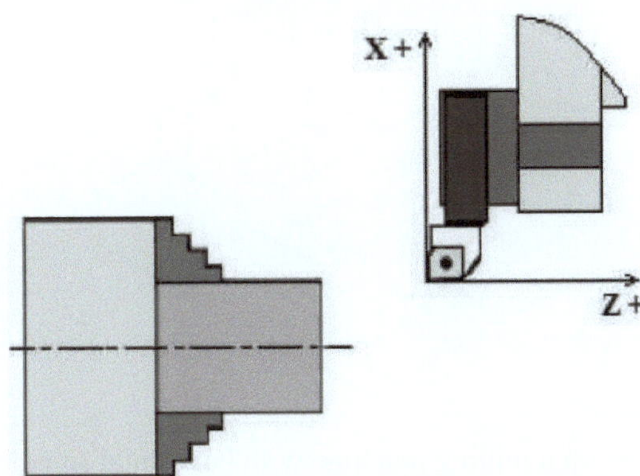

*Figure 32. Axes in the turning procedure*

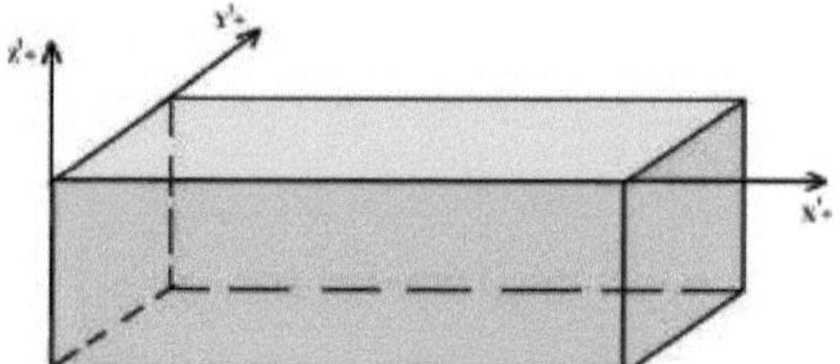

*Figure 33: Axes on a prismatic part*

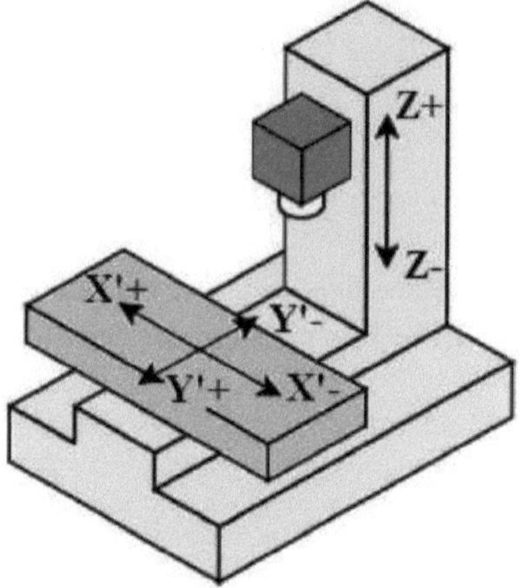

Axes on a milling machine with vertical spindle

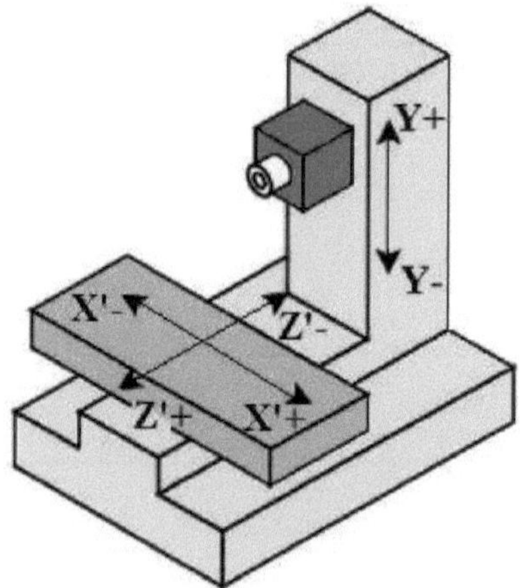

Axes on a milling machine with horizontal spindle
*Figure 34: Axis positions on NC Machines*

*Table 4. Classification by number of axes*

| Processes | Number axes | Axes | Operations |
|---|---|---|---|
| Shooting | 2 | X and Z | General turning: turning, facing, drilling, etc. |
| | 3 | X, Z and C | General turning, milling on the lathe and |
| | 4 | X, Y, Z and C | drilling off the Z axis, ... |
| | 4 | X1, X2, Z1 and Z2 | Turning using two turrets. |
| | 5 | X1, X2, Z1, Z2 and C | With the use of two turrets and milling on the lathe and drilling off the Z axis, |
| Milling | 3 | X, Y and Z | General milling: face milling, grooving, drilling, tapping and pocket milling. |
| | 4 | X, Y, Z and B | General milling with rotation of the part around the Y axis. |
| | 4 | X, Y, Z and C | General milling with rotation of the part around the Z axis. |
| | 5 | X, Y, Z, A and C | General milling and milling of left-hand surfaces. |
| | 5 | X, Y, Z, B and C | |
| | 5 | X, Y, Z, A and B | |

## 2.4 Defining offsets

$\overrightarrow{O_mOP}$ : is the vector that specifies the position of the programme origin OP relative to the measurement origin Om.

$$\overrightarrow{O_mOP} = \overrightarrow{O_mO_{pp}} + \overrightarrow{O_{pp}O_p} + \overrightarrow{O_pOP} \quad (Eq.20)$$

- $\overrightarrow{O_mO_{pp}}$ : is the vector specifying the position of the door origin part Opp in relation to the origin measures Om.

- $\overrightarrow{O_{pp}O_p}$ : is the vector that specifies the position of the part support in the workpiece carrier, this is the distance between the Op workpiece origin and the Opp workpiece carrier origin.

- $\overrightarrow{O_pOP}$ : is the vector specifying the position of the programme origin OP compared with the original part Op.

- **PREF**: is the **"Om / Opp"** part origin offset.

To avoid any ambiguity, it is necessary to specify that in filming, the PREF depends on the frequency of chuck assembly and disassembly:

- If the chuck is frequently dismantled: **PREF = Om / Opp.**

- If the chuck is never removed: **PREF = Om / Op.**

The program origin can be offset from the part origin either by setting the machine parameters or by programming.

- **DECI**: is the **"Opp / OP"** programme origin offset.

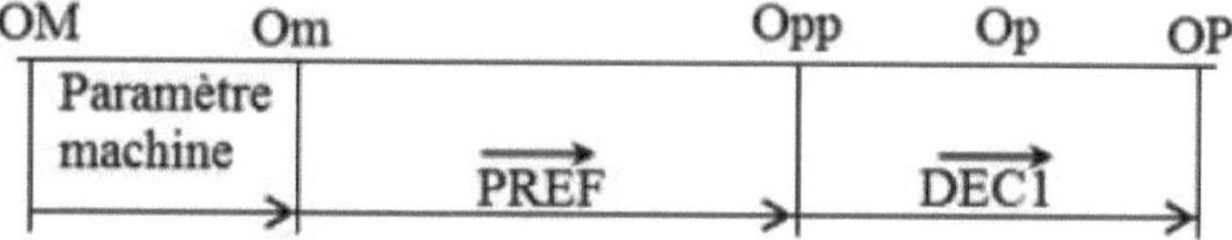

## 2.5 Machine referencing

In older CNC machines, after each start-up, the processor cannot know the position of the measurement origin in relation to the machine origin, as these machines are equipped with relative encoders. It is therefore necessary to define the "Measurement Origin" offset (ORPOM = Om/OM) after each power-down. In some machines, the measurement origin is confused with the machine origin, which means ORPOM=0.

After power-up, the operator must move the carriages in the direction defined by the manufacturer, up to the limit stops (original sensors). This enables the machine computer to determine the ORPOM values for all axes.

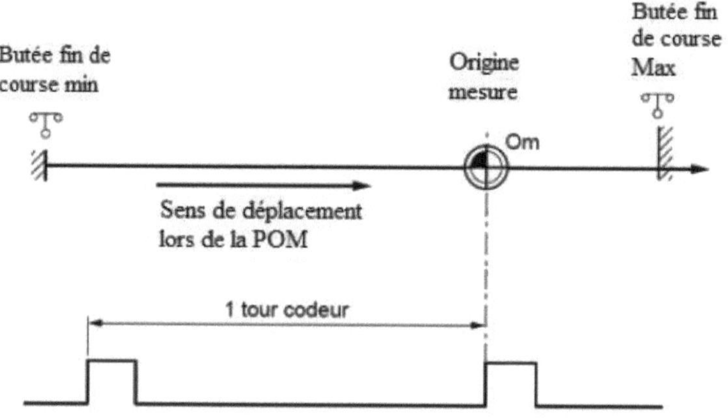


*Figure 35: Determining the measurement origin*

New NC machines are referenced automatically as soon as the machine is started up. These machines are fitted with absolute encoders. When the machine is started up, the control collects the current positions of the axes in relation to the measurement origin, without the need for a POM (Measurement Origin Pick-up).

Once the measurement origin has been located, the operator must locate the part origin, which is an accessible and known point on the part. The position can be determined either by a probe or by measuring the tangent. The tangent is defined on all axes between the point controlled by the machine and the surface of the part (it is possible to use a gauge block).

> **Milling**

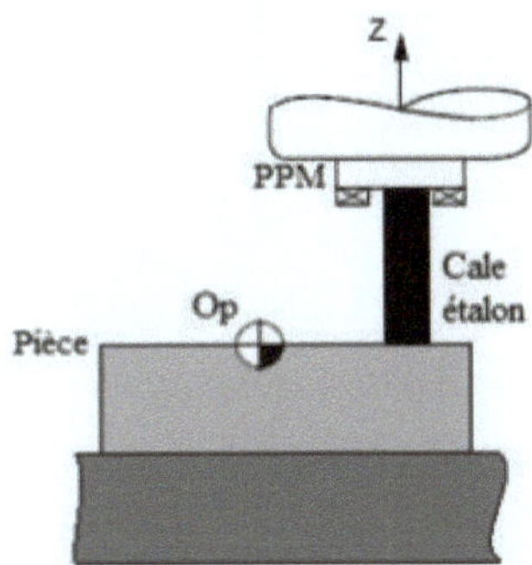



*Figure 36. Workpiece origin in relation to the Z axis*

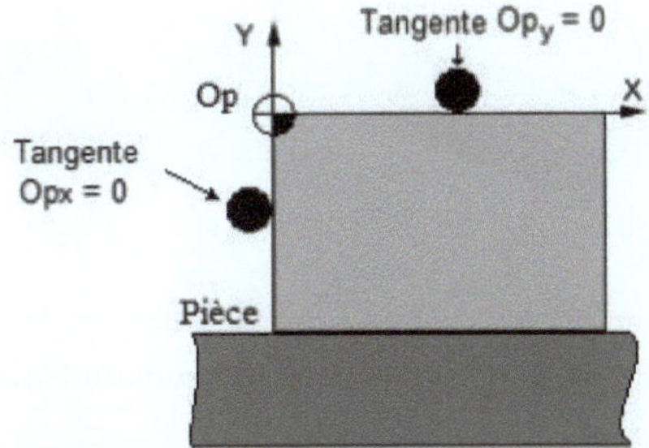



*Figure 37. Workpiece origin in relation to the X and Y axes*

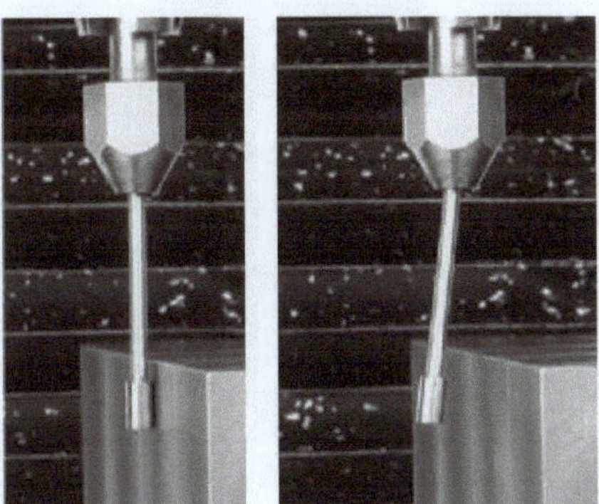

*Figure 38. Principle of part origin capture along the X axis*
When the probe touches the workpiece, an LED light comes on.

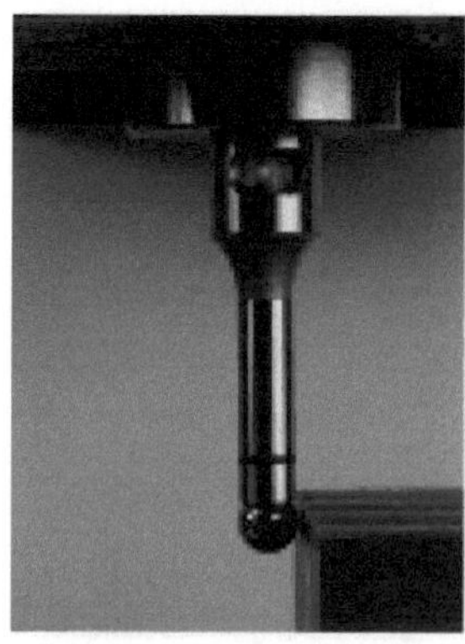

*Figure 39. Part origin capture using an electric probe with LED*

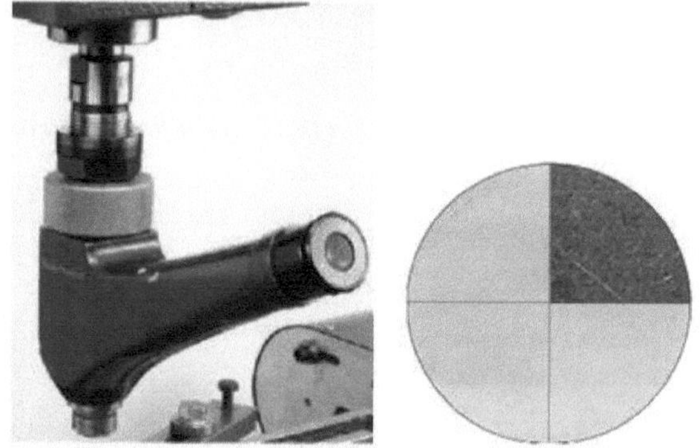

*Figure 40. workpiece pick-up an optical instrument* Enlarged view of the workpiece

*Figure 41. Original plug for a laser instrument*

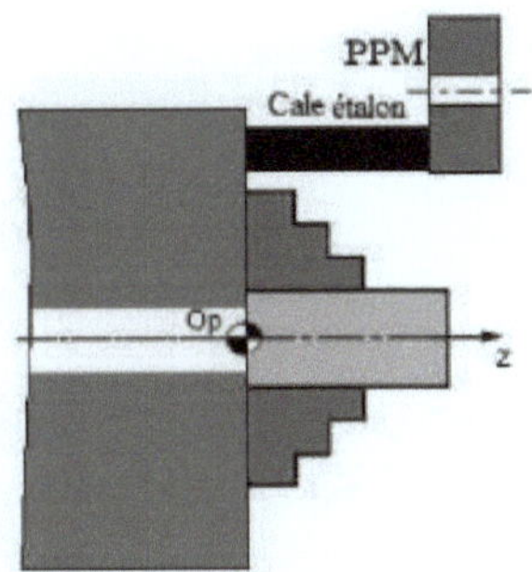



*Figure 42. Workpiece origin in relation to the Z axis*

The programmer writes the tool movement coordinates relative to the program origin (OP). These values are relative to the program origin and correspond to displacements relative to this origin. Then, when the program is executed, the machine controller converts these relative coordinates into absolute coordinates relative to the measurement origin. This conversion is essential so that the machine knows exactly where to move and positions the tool correctly on the workpiece in relation to the measurement origin, which is the machine's fixed reference.

**Coordinates relative to the measurement origin :**

$X_{Om} = x_{OP} + PREF\ X + DEC1\ X$

$Y_{Om} = y_{OP} + PREF\ Y + DEC1\ Y$

$Z_{Om} = z_{OP} + PREF\ Z + DEC1\ Z$

Obviously, when a part is fixed on a rotary table, the part origin will only be taken into account for the axis that is not affected by the rotation of the table. For rotary axes, the PREF (Reference Position) and DEC (Offset) will be determined as follows:

-    For linear axes: The co-ordinates of the workpiece origin remain unchanged as these axes do not rotate with the table.

-    For the rotary axis: The reference position (PREF) will be defined at the point where the rotary axis is at the 0 degree position (or at another selected reference angle). The offset (DEC) will be the difference between the reference position and the position of the measurement origin for this axis.

PREF = Plateau centre / Om

DEC3 = OP / Table centre: eccentricity of the part

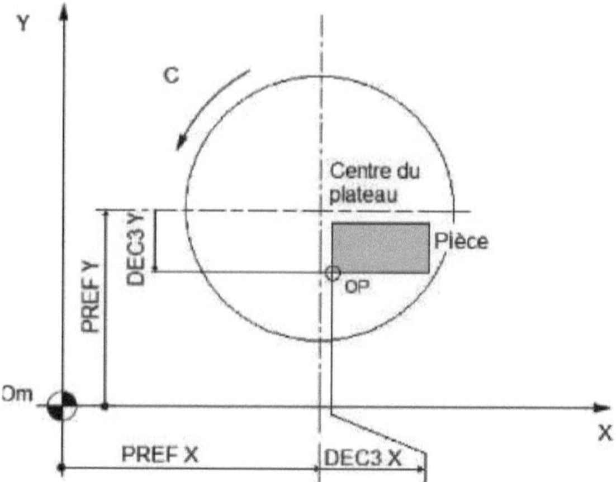


*Figure 43. Origin socket for rotary table (rotation relative to Z)*

## 2.6 Tool holders

Tool holders are intermediate elements that secure the tool to the turret when turning and to the spindle when milling. It is essential to use standardised tool holders. On the machine side, the tool holders must have the same dimensions and the same shape to allow automatic tool change, while guaranteeing good fixing and perfect stability. On the tool side, it is essential to choose a tool holder that adapts to the specific dimensions and shape of each tool.

**- On location :**

Regardless of the size of the tool, tool holders are made up of a structure that ensures the tool tip is centred in relation to the axis of the workpiece. They are chosen according to the type of turret. The most commonly used tool holders are DIN 69880 and VDI.

External radial tool holder External axial tool holder

ER collet chuck VDI drill chuck

*Figure 44. Tool holder*

## • **Milling**

Tool holders have standardised tapered shapes and are specially designed to ensure that the tool tip is centred on the axis of the workpiece. They have a central or lateral hole to allow lubricant to pass through, which is necessary for the tool to function properly and be lubricated during machining.

The most commonly used types of tool holder are :

ISO 7388/1: ISO 30, ISO 40, ISO 45 and ISO 50.

DIN 69871 : ISO 40 and ISO 50.

MAS BT: ISO 30, ISO 40 and ISO 50.

YAMAZAKI: ISO 40 and ISO 50.

DIN 2080 : ISO 40 and ISO 50.

DIN 696693: HSK 40-A, HSK 50-A, HSK 63-A, HSK 80-A and HSK 100-A

*Table 5. Milling tool holders*

| Tool holders for cylindrical end mills | |
|---|---|
| Tool holder for cylindrical end mills with flats | |
| Tool holder for centre-bored cutter with pin drive | |

## • **Drilling**

It is essential that toolholders ensure that drill tips are properly centred to guarantee hole quality. Accurate centring allows holes to be drilled with greater precision, avoiding deviations and ensuring dimensions conform to the required

55

specifications. This is particularly important in machining operations where hole accuracy is crucial to the quality of the finished product. Toolholders designed specifically for drilling play an important role in ensuring the correct alignment and stability of the drill, enabling optimum machining results to be achieved.

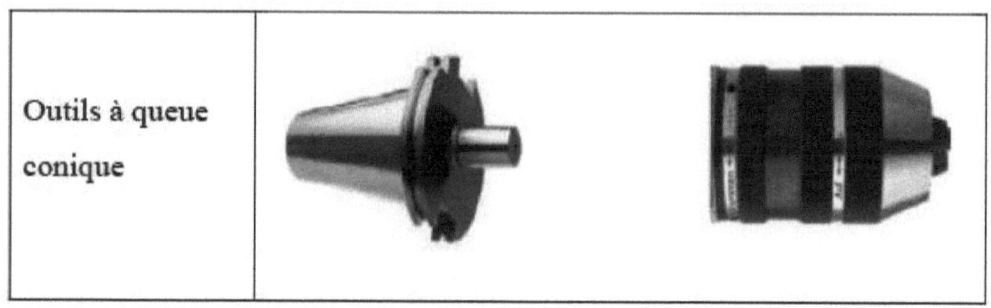


Tools with tapered shanks

## 2.7 Tool measurement

After each reassembly, it is impossible to keep the tool in exactly the same position in the tool holder. In addition, each tool has its own dimensions and shapes. The main tool types and insert shapes are shown in Appendices 1 and 2.

There are several methods for measuring tool gauges:

- With the use of a presetting bench: this solution considerably reduces machine downtime.

- Manually by the machine: this solution is only moderately effective, as it immobilises the machine during the preparation phase.

- By using measurement sensors (probes): it's an easy, quick and accurate solution".

These methods of measuring tool gauges are essential for guaranteeing the precision and quality of the machining operations carried out by the CNC machine. The choice of method will depend on the specific requirements of the machining operation and the availability of equipment in the workshop.

*Figure 45. Tool presetting benches*

<u>**Shooting:**</u> the correction parameters are :

- Tool radius.
- The dimensions of the tool in relation to the X and Z axes.
- The position of the cutting edge in relation to the centre of the circle.

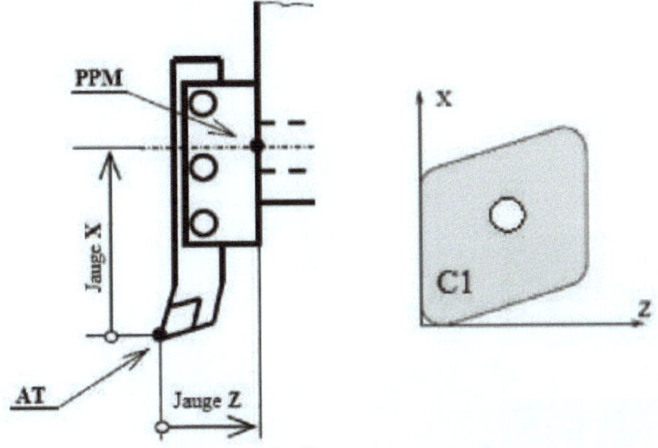


*Figure 46. Gauges in turning*

The figures below show the methods used to measure the gauges on turning tools. It consists of making a tangent between the tool and a gauge block (or a workpiece) of known length and diameter, then calculating the value of the gauges Jx and Jz.

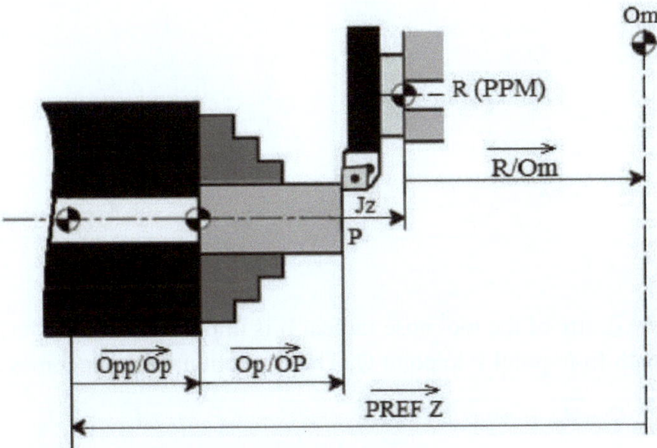


*Figure 47. Method for determining the Z gauge*

$$\overrightarrow{O_m O_{pp}} = \overrightarrow{O_{pp} O_p} + \overrightarrow{O_p OP} + \overrightarrow{OPP} + \overrightarrow{PR} + \overrightarrow{RO_m}$$

Avec :

$$\overrightarrow{PR} = \vec{J}z$$

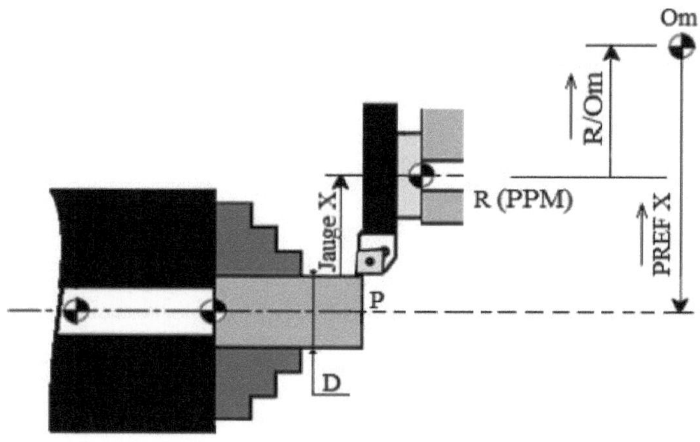

*Figure 48. Method for determining the X gauge*

With :

$$\overrightarrow{PR} = \overrightarrow{JX}$$

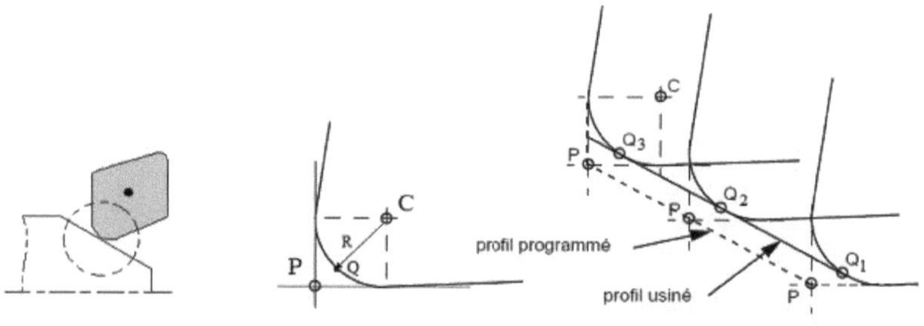


Point C is the centre of the tool nose radius. It is important to bring the machine-controlled path from point P to point Q. The correction model becomes :

$$\overrightarrow{O_m O_{pp}} = \overrightarrow{O_{pp} O_p} + \overrightarrow{O_p OP} + \overrightarrow{OPQ} + \overrightarrow{QC} + \overrightarrow{CP} + \overrightarrow{PR} + \overrightarrow{RO_m}$$

**<u>Examples:</u>**

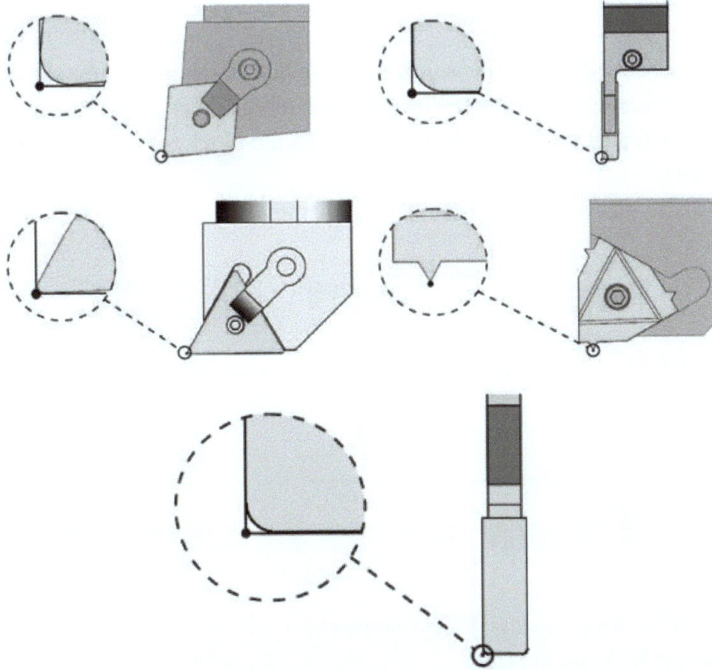

*Figure 49: Turning tool shapes*

## **Milling :**

The tool correction parameters are :
-   The length of the tool in relation to the Z axis.
-   Tool radius.

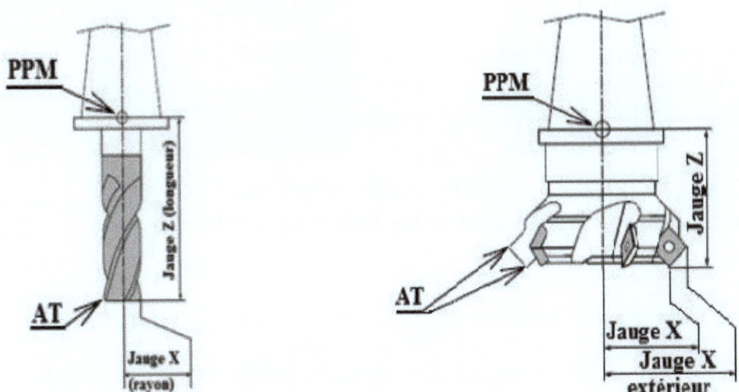


*Figure 50. Milling gauges*

There are three methods for measuring milling cutters on a CNC machine:
- Make the tangent between the tool and the table (Z=0).
- Make the tangent between the tool and a gauge block (or a piece of height). well known).
- Automatic measurement of length and diameter using a probe installed on the table.

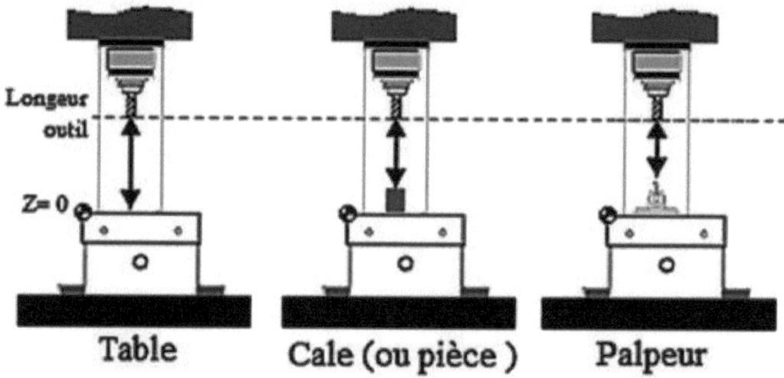


*Figure 51. Tool length measurement methods*
Automatic probe Manual probe

*Figure 52. Examples of tool length measurement*
Point P on milling tools is located in the following positions:

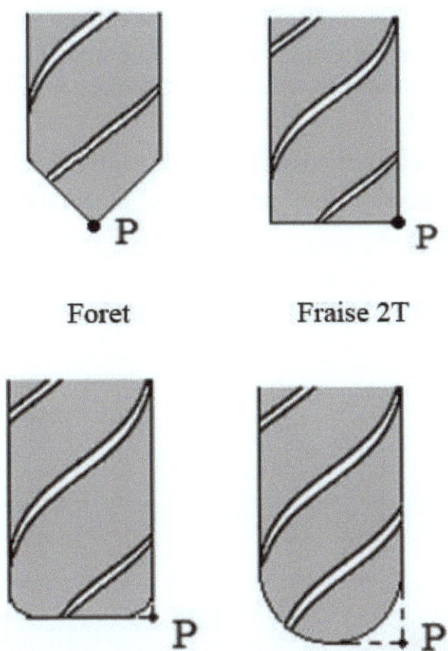

Toric milling cutter Hemispherical milling cutter
*Figure 52.   Strawberry tips*

## 2.8 CNC machine modes

Most CNC machines operate in modes that correspond to a specific use of the machine. These modes differ according to the technology, the functions available and the area of use. Here are the most commonly used modes (or sub-modes):

•   Referencing mode: this enables synchronisation between the operating part and the control part of the machine (measurement origin, current position of carriages and tool magazine).

•   JOG mode: this is a manual operating mode for setting up the machine, defining tools and the workpiece origin.

•   Programming mode: used to enter programmes manually or copy files using a USB key or connection cable.

•   Memory mode: saves programmes in the machine's memory.

•   Automatic mode: the machine operates automatically. When the operator presses the "cycle start" button, the machine executes all or part of the programme blocks.

•   IMD mode: this mode is available in FANUC machines. It enables simple operations to be programmed, tested and executed.

61

• Teach-in mode: in this mode, manual movements are memorised by the machine and converted into positions in the programme.

• Diagnosis and commissioning mode: in this mode, the operator can carry out maintenance operations".

These different modes make it possible to efficiently manage the various stages of the machining process, from programme preparation and execution to machine set-up and control.

# Chapter 3

# Programming standards

## 3.1 Introduction

To understand how numerically controlled machine tools are used to manufacture parts, you need to be familiar with the concepts of programming. Programming" in CNC refers to the codes used to transform a raw (or semi-finished) part into a finished product. This programme contains all the information needed by the machine, i.e. it describes the operations to be carried out by the operating part sequentially and using codes that can be assimilated by the operator. This coded language is called "G code".

G-code, also known as G-code programming language, is a universally used language for communicating with CNC machines. It is used to specify tool movements, speeds, coordinates and all the actions required to perform the desired machining operation. G-code programming is essential for defining the machining process, ensuring the accuracy and quality of the finished product, and optimising the efficiency of the CNC machine.

## 3.2 Evolution of programming standards

It's true that today there are more than 4,500 CNC post-processors, with the main functions generally the same in most of them. However, to improve the performance of their machines, manufacturers often add specific codes adapted to their equipment.

The CNC programming language was first standardised by the International Organisation for Standardisation (ISO), a worldwide federation of national standards bodies created in 1947 and comprising 164 member organisations. The work of this organisation has resulted in international agreements published in the form of ISO standards.

The evolution of CNC programming standards is shown in the table below:

*Table 6. Evolution of ISO programming standards (2020)*

| Standards | Publication date | Status |
|---|---|---|
| ISO 840 | 1973 | Former |
| ISO 1056 | 1975 | |
| ISO 1057 | 1973 | |
| ISO 1058 | 1973 | |
| ISO 1059 | 1973 | |
| ISO 2539 | 1974 | |
| ISO 6983-1 | 1982 | Cancelled |
| ISO 6983-1 | 2009 | Published |
| ISO 14649-1 | 2003 | Published |
| ISO 14649-10 | 2003 | Cancelled |

| ISO 14649-11 | 2003 | Updated in 2004 |
|---|---|---|
| ISO 14649-11 | 2004 | Published |

❖ The **NUM** language is based on the ISO language, but with the addition of a number of specific functions.

❖ The Deutsches Institut für Normung (**DIN**) **has** developed a programming standard for CNC machines. This standard is based on **ISO 6983**.

| **DIN 66025-1** | Digital machine control, format; general requirements |
|---|---|
| **DIN 66025-2** | Industrial automation; numerical control of machines; format, preparatory and other functions |

❖ **Electronic Industry Association (EIA)**: in the early 1960s, the EIA developed the **RS274** lagoon. Version **D** of this lagoon was standardised in 1979 under the reference ISO 6983 (RS274D). In 1992 the EIA created the **RS274/NGC** lagoon (The Next Generation Controller Part Programming Functional Specification), followed by a second version in 1994. This version has many new features compared with **RS274-D**.

❖ The **FANUC** language is also based on the ISO language, but with the addition of some specifications (for example the use of semicolons at the end of blocks).

### 3.3 General programme structure

A CNC (Numerical Control) program consists of a series of sequences (functions and addresses) that are stored in the machine's memory. When the workpiece is machined, these sequences are transformed into movements by the computer, which executes them in the programmed order. The corresponding control signals are then transmitted to the machine, resulting in the movements and actions of the tool in accordance with the CNC programme.

This process enables complex machining operations to be carried out with great precision and efficiency, following the instructions specified in the programme.

A programme mainly comprises :

❖ Preparatory address functions G.

❖ Point coordinates (X, Y, Z, I, J ...)

❖ Information on speeds (S, F ...)

❖ Auxiliary M address functions.

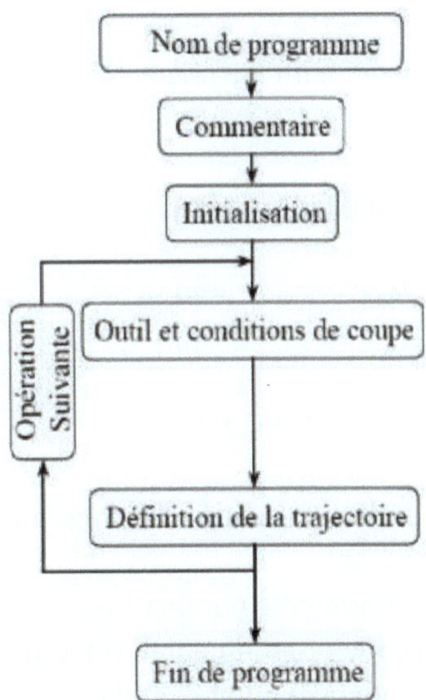


Programmes are written from left to right and top to bottom. Each sentence in a programme is called a block. Blocks are arranged and structured in a specific sequence:

1) Programme start
2) Tool call
3) Definition of cutting parameters (feed and rotation speeds)
4) Spindle rotation
5) Starting the spindle
6) Fast travel to the point of approach
7) Machining
8) Room clearance
9) Stop watering
10) Spindle stop
11) Rapid movement to a safe position
12) End of programme

**3.4 Block format**

A CNC program is made up of blocks. The format of a block is as follows:

| N. | G. | X. Y. Z. | F. S. | M. |
|---|---|---|---|---|
| Block number | Preparatory function | Position to be reached | Technological function | Auxiliary function |

In some machines, the lines end with the characters at the end of the block, for example: ";", "EOB" or "LF".

## 3.5 Word format

A word is made up of a letter called an address, for example (G, M, X, Y, etc.) and a number of digits (from 0 to 9).

| Address | Sign | Value |
|---|---|---|
| One or two letters | - or [+] | Figures |

Examples: **G01, M01, X50.3, Y-25.36, Z0.**

## 3.6 The main addresses

**%...** "Programme number": the numbering of a programme is preceded by %, the number varies from 1 to 9999.

On FANUC machines, the programme number begins with the letter "O...".

In SINUMERIK commands, the program name is the file name.

**N...** "Line number": a chronological number at the beginning of the line, which can take values from 0 to 9999. The numbering does not affect the order of execution and is not compulsory. For ease of reading, lines are numbered in increments of 10.

**G...** "Preparatory functions" defining the form and conditions of travel.

**M...** "Auxiliary functions" indicate changes in machine status.

**X... Y... Z...** "Main axes" designating the coordinates of the arrival points.

**I... J... K...** Parameters defining circular trajectories (centre position).

**R. ..** Parameters defining circular paths (radius).

**S. ..** Specifies the spindle speed.

**F...** Specify the feed speed.

**T. ..** Tool number symbol.

**D...** Tool corrector.

**A. ..** Axis of rotation around X.

**B. ..** Axis of rotation around Y.

**C.** Axis of rotation around Z.

**U., V.** and **W.** Secondary axes.

**P.** and **Q.** used in some machining cycles.

## 3.7 Preparatory functions

### 3.7.1 Rating systems

In the following, we will concentrate mainly on ISO standard functions. The programmed dimensions can be expressed in the following ways:

| Description | ISO |
|---|---|
| Absolute programming: the dimension is referenced to the programme | G90 |

| origin. | |
| Relative programming: the dimension is referenced to the previous tool position. | G91 |
| Absolute programming in measurement origin: the dimension is referenced to the machine's measurement origin "om". | G52 |

**Syntax :**

> G90 /G91 /G52/G500/G53.......

## Example 1

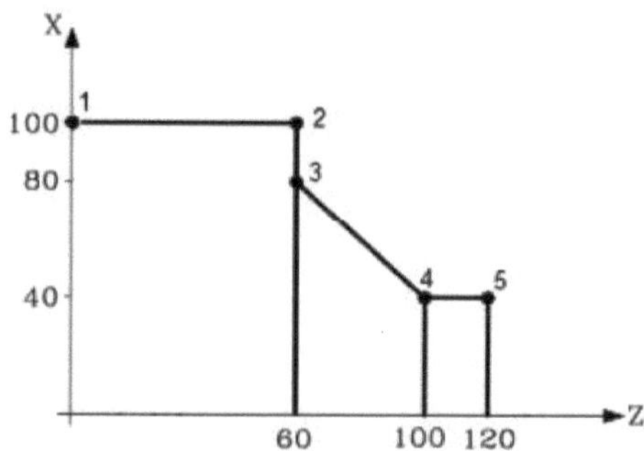

## Correction to example 1

| Absolute programming | Relative programming |
|---|---|
| X0    Z0 ; Origine | X0    Z0 ; Origine |
| G90   X100 Z0   ; Point 1 | G91   X100 Z0  ; Point 1 |
| X100 Z60  ; Point 2 | X0    Z60 ; Point 2 |
| X80   Z60  ; Point 3 | X-20 Z0 ; Point 3 |
| X40   Z100 ; Point 4 | X-40 Z40 ; Point 4 |
| X40   Z120 ; Point 5 | X0    Z20 ; Point 5 |

## Example 2

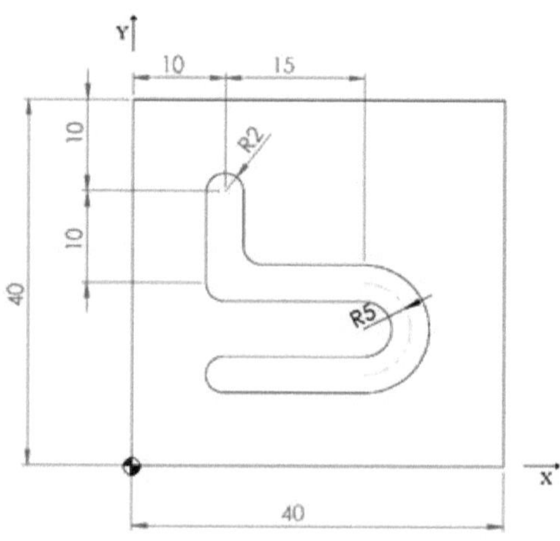



**Correction of example 2**

**Absolute programming Relative programming**

| | |
|---|---|
| G90 | G00 X10 Y10 |
| G0 X10 Y10 | G91 X15 F80 |
| X25 | G3 X0 Y10 R5 |
| G3 X25 Y20 R5 | G1 X-15 |
| G1 X10 | Y10 |
| Y30 | |

## 3.7.2 Coordinate systems

| Description | ISO |
|---|---|
| Moving in Cartesian coordinates | G21 |
| Moving in polar coordinates | G20 |
| Displacement in cylindrical coordinates | G22 |

## 3.7.3 Measurement systems

| Description | ISO |
|---|---|
| Activating the **metric** system | G71 |
| Activation of the measurement system in **inches** | G70 |

## 3.7.4 Scale factor

| Description | ISO |
|---|---|
| Activation of the **scale** factor | G74 |
| Deactivation of the **scale** factor | G73 |

**Syntax :**

G74 E69000 = .....

The factor is between 1/1000 and 9999/1000 and must be a whole number.
Function G74 is modal and must not be used with functions G02, G03, G41 and G42.

**Examples:**
G74 E69000 = 3000 (here a multiplication of 3)
G74 E69000 = 500 (here a reduction of 1/2)

## 3.7.5 Mirror function

This function is used to machine a programme symmetrically.

| Description | ISO |
|---|---|
| Activating the mirror function | G51 |
| Deactivating the mirror function | |

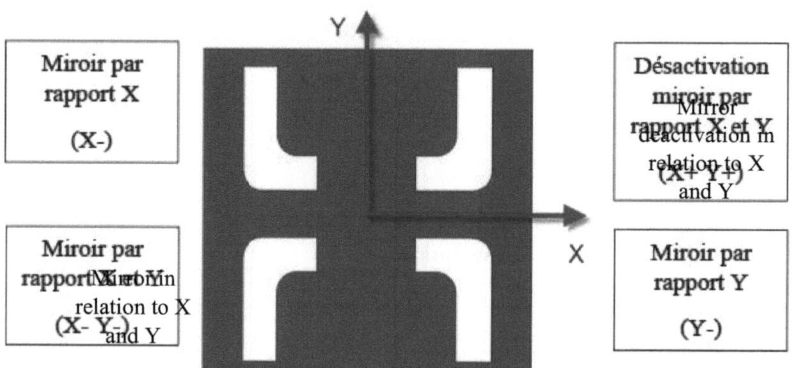


<u>Syntax :</u>

G51 X- Y- Z-

The negative sign (-) validates the mirror in relation to one or more axes.

The G51 function followed by one or more arguments X+, Y+, Z+ revokes the previous G51 state.

Function G51 must be programmed alone in a block.

**Examples:**

G51 X- Y- (in this case symmetry about the X and Y axes)

G51 X+ (in this case cancelling the symmetry about the X axis)

## 3.7.6 Original offset

This is a programmed offset (programme origin) which is used to designate programme origins in several parts of the part. This function is useful for repetitive shapes or if several identical parts are assembled on the same machine.

| Description | ISO |
|---|---|
| Activating the origin offset | G59 |
| Deactivating the zero offset | |

<u>Syntax :</u>

*X...*, *Y.* and *Z.* are the coordinates of the new program origin.

G59 X0 Y0 Z0: deactivates the zero offset.

**Example**

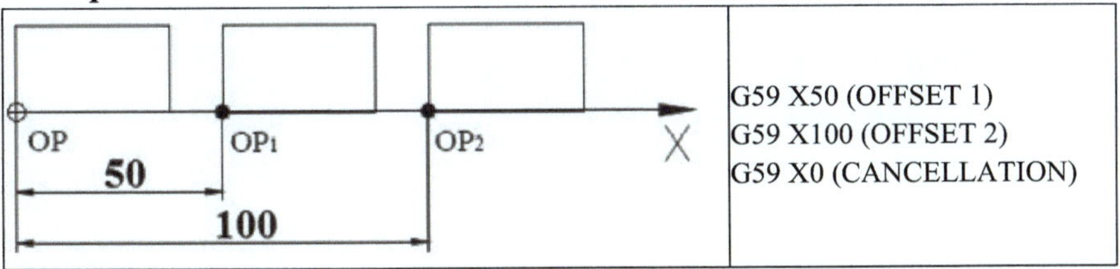



## 3.7.7 Tools

### 3.7.7.1 Tool call with offset

<u>Syntax :</u>

The method of calling up a tool with its corrector varies from one machine to another:

```
T... D... M6
```

With :

**T..** : the tool number (example T1)

**D..** : the number of the tool spacer (example D1)

**M6**: tool change order

### 3.7.7.2 Tool radius correction

The point controlled by the machine corresponds to the axis of the milling cutter in milling and to the tool holder in turning. On the other hand, the point of the cutting edge is on the periphery of the milling cutter and on the tip of the tool in turning.

To make programming easier, especially in contouring operations, it is recommended to program in relation to the machine controlled point (PPM) and to use one of the correction functions to offset the tool by a value equal to the tool radius. This approach allows operations to be carried out more intuitively, based on the centre of the tool, which simplifies path programming and guarantees more accurate results during machining.

**G41**: Correction of the radius on the left of the profile to be machined

**G42**: Radius correction on the right of the profile to be machined

**G40**: Cancels the correction.

So the program includes the coordinates of the workpiece profile, and the machine controller calculates the position of the cutting edge.

<u>**Correction for milling :**</u>

G41          G40          G42

## Correction for filming :

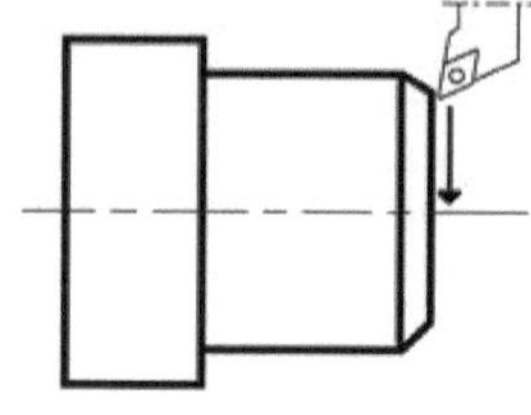
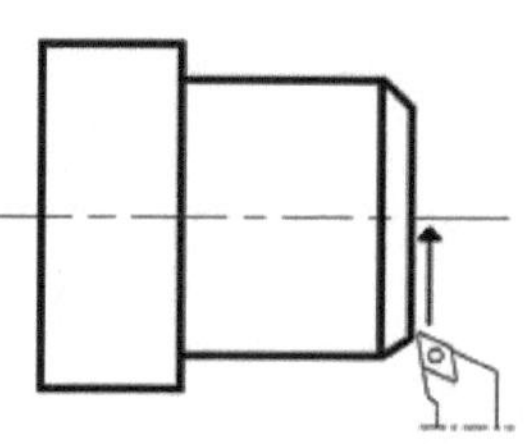

**G41**          **G42**

## Syntax :

> [G00/G01/G02/G03] G41/G42 …

## 3.7.8 Cutting conditions
### Syntax :

Feed speed

> G94/G95 F….

**G94**: Feed speed (mm/min), for milling and turning.
**G95**: Feed rate (mm/rev), for milling and turning.

Cutting speed

> G96 S….

**G96:** Constant cutting speed in (m/min), when turning.
- It is advisable to turn the spindle using G97 before calling up function G96,
- Before changing a tool, it is advisable to revoke the cutting speed by defining the rotation speed using function G97.

Speed of rotation

> G97 S….

**G97**: Speed of rotation (rpm), for milling and turning.

Limitation of rotation speed

G92 S….

**G92**: Limitation of rotation speed (rpm), when turning.

**Note:** for FANUC type machines

**G50**: Limitation of rotation speed (rpm) for models A and B of CNC programmes.

**G92**: Speed limitation (rpm) for model C CNC programmes.

**Examples:**

G94 F100: Feed speed f = 100 mm/min.

G95 F0.5: Feed speed f= 0.5 mm/rev.

G96 S23: Cutting speed Vc= 23 m/min.

G97 S1500: Rotation speed N = 1500 rpm.

G92 S9000 : Maximum spindle speed N = 9000 rpm.

**3.7.9 Linear interpolation**

**Syntax** :

G00/G01 X … Y … Z…

**G00**: Linear interpolation at high speed.

**G01**: Linear interpolation at programmed speed.

**X, Y** and **Z**: are the coordinates of the arrival point.

**Examples:**

**For the shoot :**

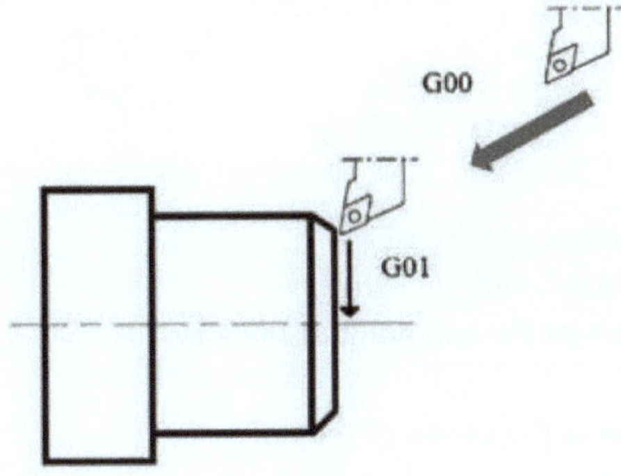


**For milling :**

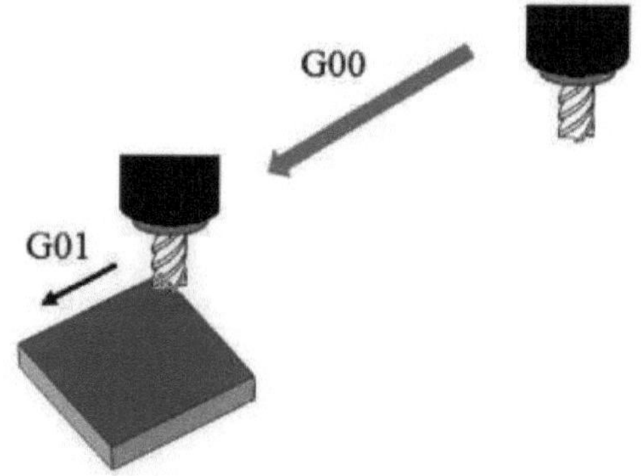


### 3.7.10Circular interpolation
**Syntax** :

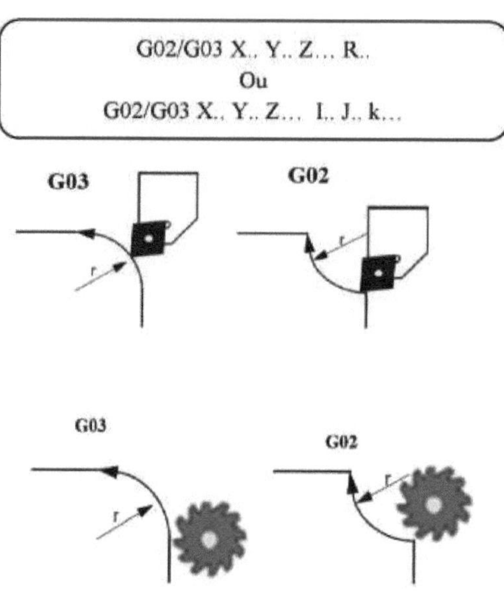


**G02**: Circular interpolation, clockwise.

**G03**: Circular interpolation, anticlockwise.

**X, Y** and **Z**: Coordinates of the end point of the arc

**R**: Arc radius

**I, J** and **K**: Coordinates of the centre of the arc

I next X.

J suivant Y.

K following Z.

### ❖ Shooting:

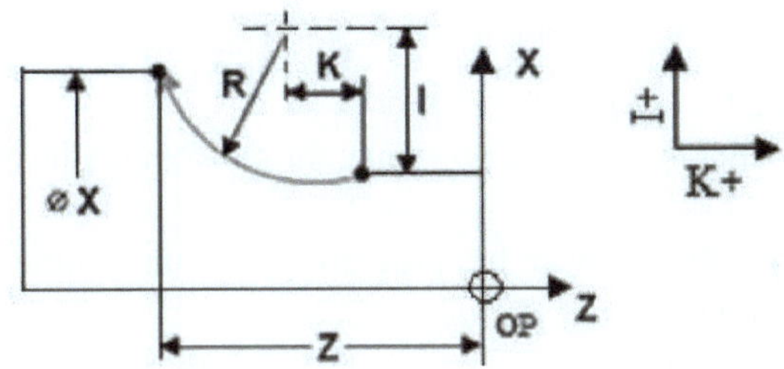


### ❖ Milling:

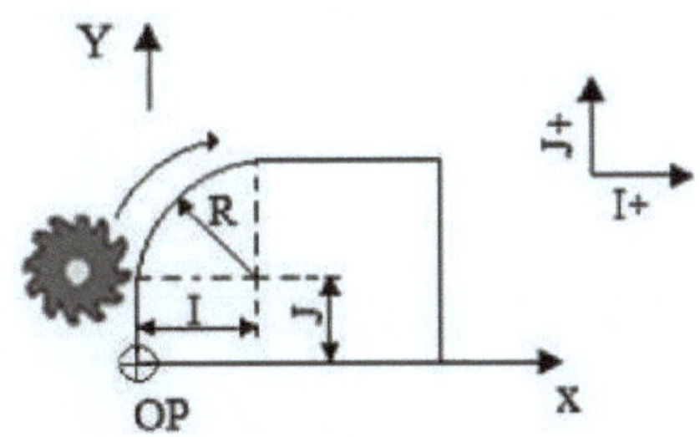


## Example 1 shooting :

Perform the contouring operation on the following part:

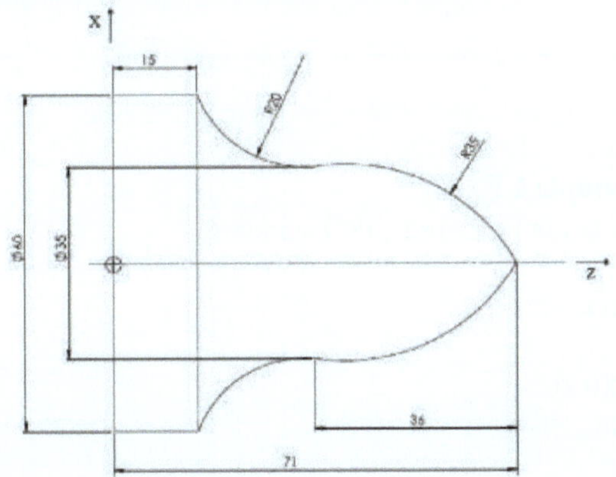


## Correction of example 1 :

```
%100
N10G0        X0 Z75
N20M03       S200
    N30G01Z71F200
    N40G03X35Z35R35
    N50G02X60Z15R5
    N60G01Z0
```

N70X62
N80M02

## Example 2 milling :

For the part shown below, give the programs for performing the contouring operation.

1. Circular interpolations must be programmed with centre coordinates in accordance with the ISO standard.

2. Circular interpolations must be programmed with radii in accordance with the ISO standard.

**Tool:** Two-size milling cutter **d= 8 mm**; **T4D4**; **Vc = 26 m/min**; **f = 100 mm/min**.

The approach point for the contouring operation is (**X= 0, Y= -20, Z= -5**).

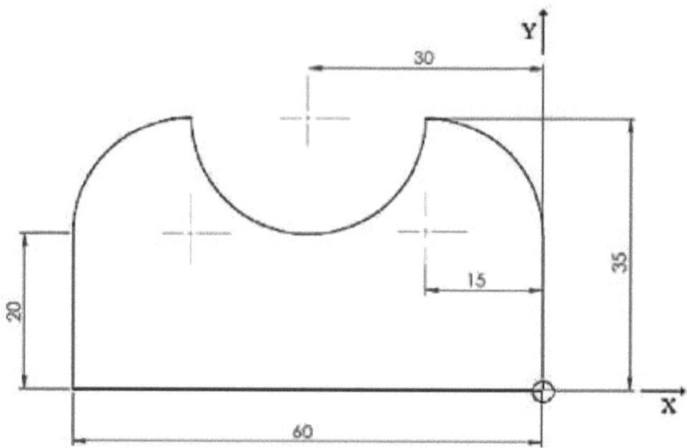


## Correction of example 2 :

1- Circular interpolation by centres, ISO standard :

%1234

N10G80      G90 G40 G71

(Tool call)

N20G52      G00 X0 Z0

N30T4       D4 M6

(Cutting conditions)

N40G97      S1000 M3 M41

N50G96      S26

N60G94      F100

(Contouring operation)

N70     G00 X 0 Y-20, Z-5 G42

N80     G01 Z110 M3 M7

N90     Y20

N100    G3 X-15 Y35 I-15 J20

N110    G2 X-45 Y35 I-15 J30
N120    G3 X-60 Y20 I-15 J20
N130    G1 Y0
N140    G1 X10
N150    G97 S1000 M9
N160    G52 G40 G00 X0 Z0 M5
N170    M02


2- Circular radius interpolation, ISO standard :
        %1234
N10     G80 G90 G40 G71
        (Tool call)
N20     G52 G00 X0 Z0
N30     T4 D4 M6
N40     (Cutting conditions) G97 S1000 M41
N50     G96 S26
N60     G94 F100
N70     (Contouring operation) G00 X 0 Y-20, Z-5 G42
N80     G01 Z110 M3 M7
N90     Y20
N100    G3 X-15 Y35 R15
N110    G2 X-45 Y35
N120    G3 X-60 Y20
N130    G1 Y0
N140    G1 X10
N150    G97 S1000 M9
N160    G52 G40 G00 X0 Z0 M5
N170    M02

### 3.7.11 Time delay

This function suspends programme execution for a programmed time. Function G04 is not modal.

**<u>Syntax :</u>**

```
G04 F...
```

**G04**: Programmable timer function.

**F...** Time in seconds from 0.01 to 99.99 seconds.

Note: function "G04 F.." does not cancel the feedrate value programmed with F

Example:

G04 F6.5 (6.5 second delay).

### 1.1.12 Stop at end of block

This function stops execution of the programme at the end of the current block.

**Syntax :**

G09

### 1.1.13 Machining plans

This function is used to select the plane for circular interpolation and ray correction.

**Syntax :**

G17/G18/G19

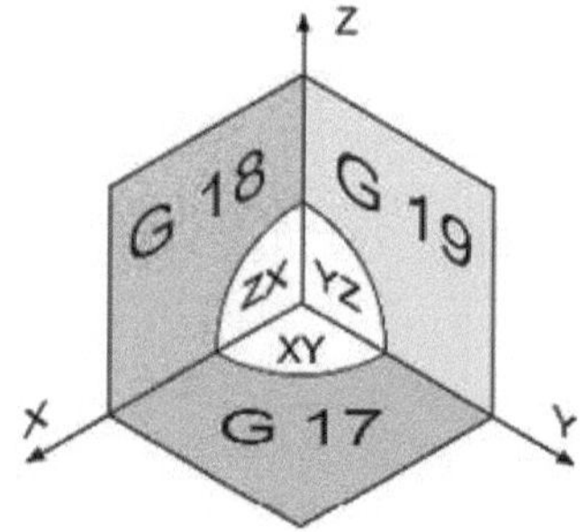



**G17**: XY (or UV) plane.
**G18**: ZX (or WU) plan.
**G19**: YZ (or VW) plan.

**Remarks :**

Function G17 is initialised when the machine is switched on.

G17, G18 and G19 are modal functions.

### 3.8 Auxiliary functions

### 3.8.1 Programmed stop

**Syntax :**

M03/M04/M05

The "M00" function stops execution of the programme in progress. The cycle will be restarted (next block) when the operator presses the cycle start button, depending on the specific operation of the machine.

The "M01" function also stops execution of the programme in progress, but only if the optional stop button is activated. The next block will be executed when the operator presses the cycle start button.

The "M02" function is used to declare the end of programs. It is used to end programme execution. If the operator presses the cycle start button after the "M02" function, the programme will be executed starting with the first block.

Finally, the "M30" function declares the end of the programme, accompanied by a machine reset. This function is often used to indicate that the programme has finished and that it is time to prepare the machine for a new operation or programme.

### 3.8.2 Spindle rotation

**Syntax :**

M03/M04/M05

**M03**: Clockwise spindle rotation

**M04**: Spindle rotation in trigonometric direction

**M05**: Spindle stop

### 3.8.3 Watering

Functions M07, M08 and M09 control the sprinkler pumps.

**Syntax :**

M07/M08/M09

**M07**: Activates first watering.

**M08**: Activates second watering.

**M09**: Watering deactivation (first and second).

Functions M07 and M08 are modal functions and are revoked by either M09 or M02.

The M09 function is a modal function initialised when the machine is switched on.

### 3.8.4 Speed ranges

These functions can be used to define spindle rotation ranges.

**Syntax :**

M40/M41/M42/M43/M44/M45

Functions M40 to M45 are modal functions.

The minimum and maximum speeds for each range are preset by the machine manufacturer. Example for NUM 1060M machines:

M40 = 50 to 500 rpm

M41 = 400 to 900 rpm

M42 = 800 to 4200 rpm

### 3.8.5 Potentiometer control

These functions can be used to authorise or prohibit the use of the feed and rotation potentiometers while a program is running.

**Syntax :**

M48/M49

**M48**: Enables feed and spindle potentiometers in automatic mode.

**M49**: Deactivates feed and spindle potentiometers in automatic mode.

### 3.8.6 Angular positioning of the spindle

The M19 function allows the spindle to be positioned at a precise angular position, which can be useful for threading, tapping, reaming, etc. operations.

**Syntax :**

[S..] [M03/M04] [M40 ... M45] EC±.. M19

**S**: Spindle speed in rpm.

**M03/M04**: Direction of rotation.

**M40 ... M45** : Spindle rotation ranges

**EC±:** Indexing angular value in degrees.

**M19**: Spindle indexing (positioning).

**Note:** M19 is a modal function before decoding; it is revoked by functions M03, M04 and M05.

### 3.9 Sub-programs

The principle consists of dividing the programme into a main programme (PP) and one or more sub-programmes (SP). Once a sub-program has been executed, the machine controller resumes execution of the next block of the main program. It is also possible to call a second sub-program from a sub-program already running. This modular programming method simplifies code management, reuses common sequences of operations and makes programming more efficient and easier to maintain. Sub-programs are often used to handle repetitive tasks or complex machining sequences that can be called from different places in the main program, contributing to more structured and clear programming.

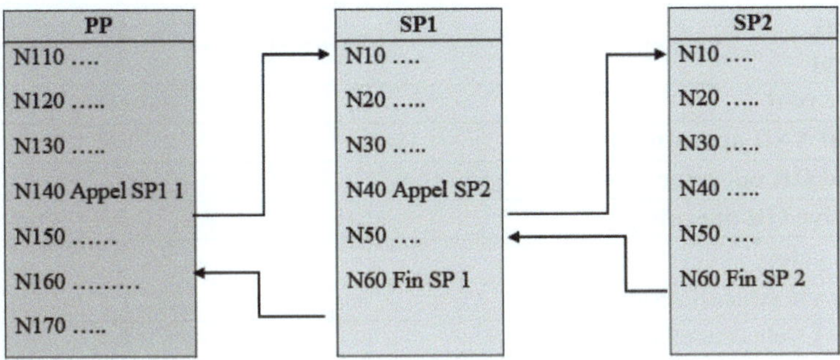


<u>**Syntax :**</u>

❖ **NUM machines :**

$$G77\ H...\ S...$$
$$ou$$
$$G77\ N...\ N...\ S...$$

**Characters recognised by post-processors**

The main characters recognised by NC machine post-processors are shown in the table below.

*Table 8. NC program characters*

| Characters | ISO |
|---|---|
| Letters of the alphabet | A ... Z |
| Figures | from 0 to 9 |
| Start of programme | % |
| Comment | ( ) |
| Decimal point | . |
| Addition sign | + |
| Subtraction sign | - |
| Multiplication sign | * |
| Sign of division | / |
| Equal (assignment) | = |
| Superior | > |
| Lower | < |
| Greater than or equal to | > = |
| Less than or equal to | <= |
| Different | < > |
| Equal to (comparison) | = |
| Absolute value | |
| Sinus | S |

| Cosinus | C |
|---|---|
| Tangent | T |
| Square root | R |
| Logical AND operator | & |
| Logical OR operator | ! |
| Exclusive OR operator | |

## 3.10  Cycles

**G80**: Cycle cancellation.

| G80 | Cycle cancellation |
|---|---|
| G81 | Centring drilling cycle |
| G82 | Drilling cycle versus reaming cycle. |
| G73 | Drilling cycle with chip breaker |
| G83 | Drilling cycle with deburring |
| G74 | Left-hand tapping cycle |
| G84 | Right-hand tapping cycle. |
| G85 | Boring cycle |
| G71 | Axial roughing cycle |
| G72 | Radial roughing cycle. |
| G73 | Roughing cycle following a profile. |
| G75 | Groove roughing cycle |
| G76 | Thread turning cycle |
| G64 | Contouring cycle |

**Exercises**

**Exercise 1**

Place the origins and offsets on the figures below:

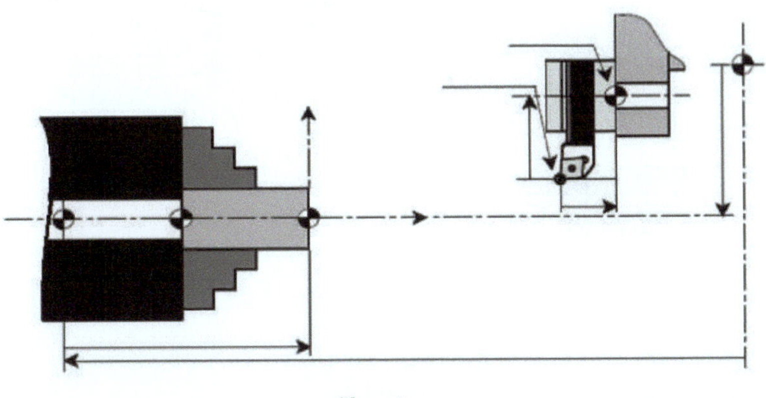

*Shooting*

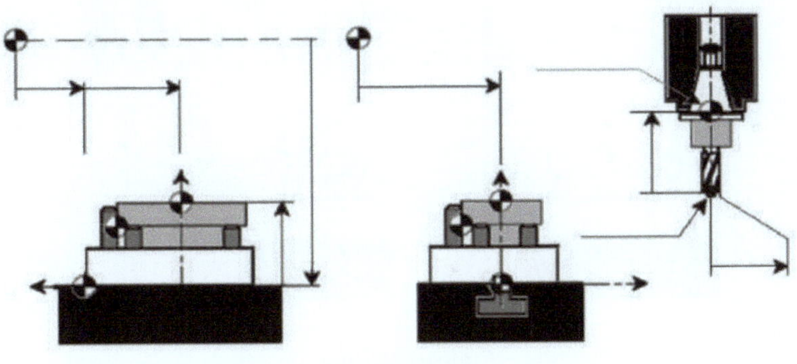

*Milling*

## Correction exercise 1

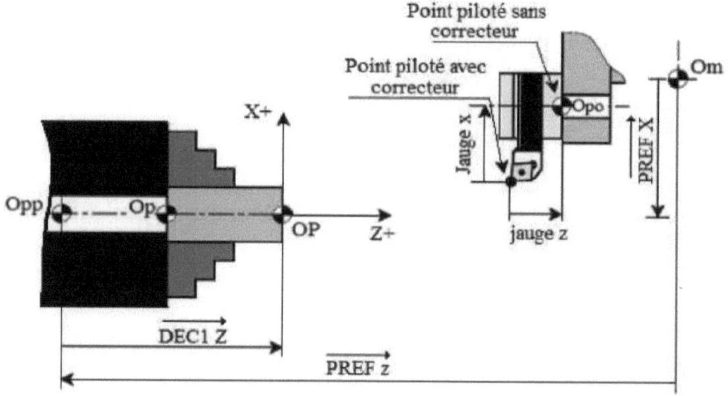

*Figure: Turning*

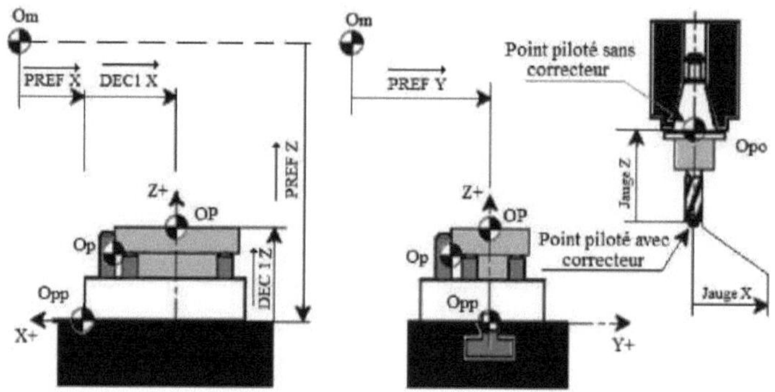

*Figure: Milling*

## Exercise 2

Or to finish machine the outer contour of the part shown below.

We give you :

- T3D3 roughing tool VC1 = 50 m/min f1 = 0.1 mm/rev.
- The approach point is 2 mm from the workpiece.
- The clearance point is 1 mm from the workpiece.

**a-** Give the name of the programme.

**b-** Initialise the machine.

**c-** Call up the tool with the corrector.

**d-** Enter the cutting conditions.

**e-** Perform the contouring operation (approach, machining and clearance).

**f-** Give the instructions for ending the programme.

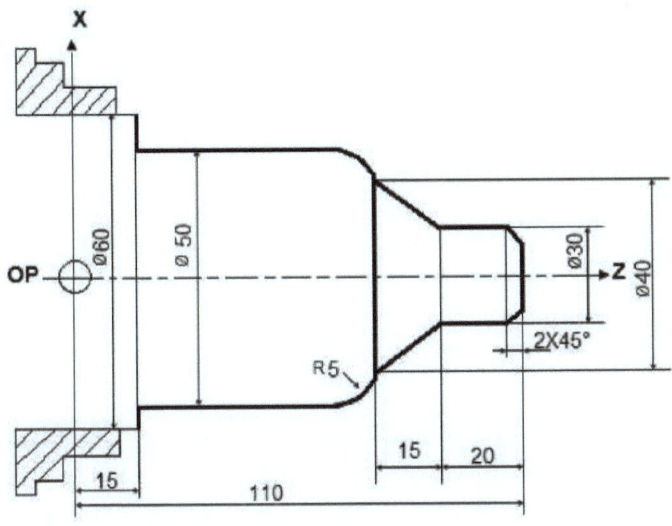

**Correction exercise 2 (ISO standard)**

% 0002

( TD2 ISO correction) (Initialisation)

N10G80     G90 G40 G71 G92 S3000

(Tool call)

N20G52     G00 X0 Z0

N30 T3 D3 M6

(Cutting conditions)

N40G97     S1000 M41

N50G96     S50

N60G95     F0.1

N70 (Contouring operation)

N80G00     X0 Z112 G42

N90G01     Z110 M3 M7

   N100X26

N110X30     Z108

   N120Z90

N130X40     Z75

N140G03     X50 Z70 R5

N150G01     Z15

   N160X62

(end of programme)

N170G97    S1000 M9
N180G52    G40 G00 X0 Z0 M5
    N190M02

Exercise 3

Finish the bore :

- boring tool T30 D30 Vc = 40 m/min; f = 0.2 mm/rev.
- Approach and clearance distance 3 mm.

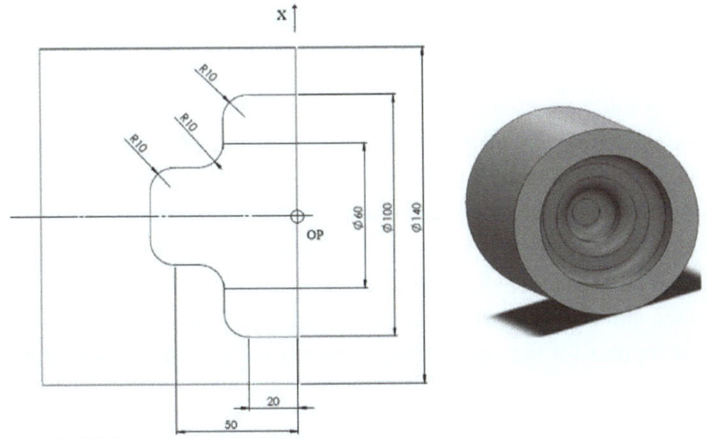

## Correction exercise 3

% 0003

(Bore)

N10G80      G90 G40 G71 G92 S3000

N20G52      G00 X0 Z0

N30T30      D30 M6

   N40G97S800M41

   N50G96S40

   N60G95F0      .2

   N70G00X100Z3G41

N80G01      Z-20 M3 M7

N90G03      X80 Z-30 R10

N100G01     X60

N110G02     X40 Z-40 R10

N120G01     Z-50

N130G03     X20 Z-60 R10

N140G01     X0

N150 Z3

N160G97     S800

N170G52     G00 X0 Z0 M9 M5

   N180M02

**Exercise 4**

Given the part shown below, give a program for finishing :

1- The outer contour.
2- The two centring operations.
3- The two drilling operations.
4- The lamage operation.
5- The tapping operation.

The following tools are available:

- Two-size cutter **d= 24 mm; T4D4; Vc = 26 m/min; f = 100 mm/min.**
- Centring drill **d= 4 mm; T1 D1; Vc = 23 m/min; f=105 mm/min.**
- Drill **d= 5 mm; T6D6; Vc = 20 m/min; f = 90 mm/min.**
- Countersink **d=12 mm; T7D7; Vc = 20 m/min; f = 120 mm/min**
- **M6** tap, **T20 D20; Vc = 8 m/min; pitch = 1 mm** and **EK=1**.

The approach and clearance distances are 3 mm from the workpiece.

The approach point for the contouring operation is (**X= -50, Y= -30, Z= - 5**).

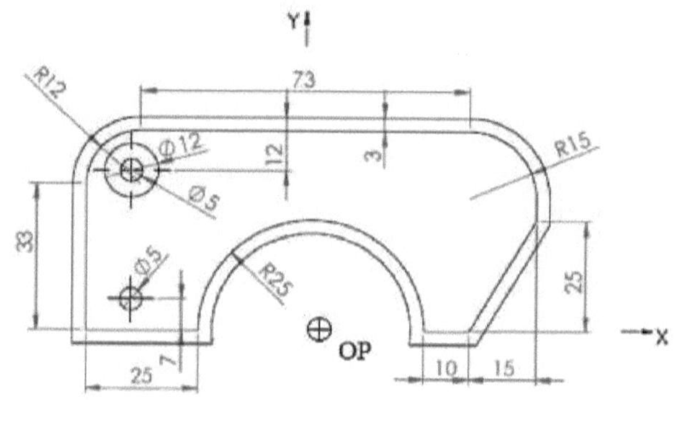


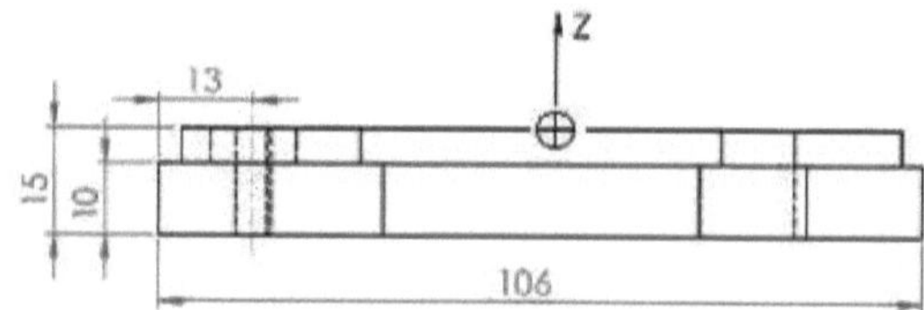

**Correction exercise 4**

%  2016        (Drilling)

( Contouring)

N10 G80 G90 G40 G71

N20 G52 G00 X0 Y0 Z0

N30 T4 D4 M6

N40 G97 S800 M41

N50 G96 S26

N60 G94 F100

N70 G00 X-50 Y-30 Z-5

N80 G41

N90 G01 Y33 M3 M7

N100 G02 X-38 Y45 R12

N110 G01 X23

N120 G02 X50 Y25 R15

N130 X35 Y0

N140 X25

N150 G03 X-25 R25

N160 G01 X-53

N170 G97 S800 M9

   G52 G40 G00 X0 Z0

N180 M5

N190 (Centering)

N200 T1 D1 M6

N210 G97 S800 M41

N220 G96 S23

N230 G94 F105

N240 G00 X-40 Y7 Z3

250 G01 Z-4 M3 M7

N260 Z3

N270 Y36

N280 Z-4

N290 Z3

G97 S800 M9

G52 G00 X0 Z0 M5

---

N300 T6 D6 M6

N310 G97 S800 M41

N320 G96 S20

N330 G94 F90

N340 G00 X-40 Y7 Z3

N350 M3 M7

N360 G81 Z-18 ER3

N370 Y36 Z-18

N380 G80

N390 G97 S800 M9

N400 G52 G00 X0 Z0 M9 M5

(Lamage)

N410 T7 D7 M6

N420 G97 S800

N430 G96 S20

N440 G94 F120

N450 G00 X-40 Y36 Z3

N470 G01 Z-5 M3 M7

N480 Z3

N490 G97 S800 M9

N500 G52 G00 X0 Z0 M5

(Tapping)

N600 T20 D20 M6

N610 G97 S800 M41

N620 G96 S8

N630 G00 X-40 Y7 Z3

N640 G84 Z-18 ER3 K1 M3 M7

N650 G97 S800 M9

N660 G52 G40 G00 X0 Z0 M5

N670 M02

Bibliography

**Overages**

[1] Christian Rattat: CNC milling for makers, Dpunkt Verlag and Rocky Nook, (2017).

[2] Peter Smid: CNC control Setup for Milling and Turning, Industrial Press, Inc, (2010).

[3] Peter Smid: CNC programming handbook, Industrial Press, Inc (2007).

[4] Marion Sabourdy and Jean-Michel Molenaar: Les machines à commande numérique: Découpeuse, fraiseuses, imprimantes 3D, Eyrolles, (2018).

[5] Kip Hanson: Machining For Dummies, Dummies, (2017).

[6] Harold Hall: Metal Lathe for Home Machinists, Fox Chapel Publishing, (2016).

[7] J-P. URSO: Commande numérique programmation, Mémotech, (2002).

[8] Jean-Pierre Urso: Mémotech - Commande Numérique - Programmation, Educalivre, (1999).

[9] R Cameron: Technology and CNC machining, Saint-Martin, (1996).

[10] C. Marty: La Pratique de la Commande Numérique des Machines-Outils Technique et Documentation, Lavoisier, (1993).

[11] Philippe Marin, Claude Marty and Claude Cassangers: La pratique de la commande numérique des machines-outils, Tec & Doc, (1999).

[12] Binding - January: Les Techniques de commande numérique des machines- outils. Monographies du centre d'actualisation Scientifique et technique, Masson Et Cie, (1970).

[13] Bernard Méry, Machines à commande numérique : De l'étude des structures à la maîtrise du langage, Hermes Sciences, (1997).

**Standards :**

[14] Automation systems and integration - Numerical control of machines - Program format and definition of address words ISO 6983-1:2009.

[15] Industrial automation systems and integration-Physical device control-Data model for computerized numerical controllers ISO 14649-1:2004.

**Machine manuals :**

[16] Programming manual, NUM 1020/1040/1060M.

## Appendix 1: Tools

**1- Shooting**

**Designation Photo**

| | |
|---|---|
| Straightening tool<br>Trolley tool<br>Carving and straightening tool<br>Copy tool<br>Boring tool<br>Tapping tool<br>Threading tool<br>Cutting tool | |

## 2- Milling
**Designation Photo**

| | |
|---|---|
| 2T face milling cutter<br>Face milling cutter<br>2T tapered shank milling cutter 2T bore and pin drive milling cutter<br>2T milling cutter with cylindrical shank<br>2T two-lip centre-cut groove cutter<br>3T milling cutter with alternating teeth<br>Extendable 3T milling cutter with alternating teeth |  |

| | |
|---|---|
| T-slot cutter<br>Type A conical milling cutters<br>Type B conical milling cutters<br>Milling cutter for gears or racks<br>Convex cutter<br>Carbide-tipped milling cutter | |

## 3- Drilling
**Designation Photo**

| | |
|---|---|
| Drill<br>Step drill<br>Centre drill<br>Machine reamer<br>Machine tap | 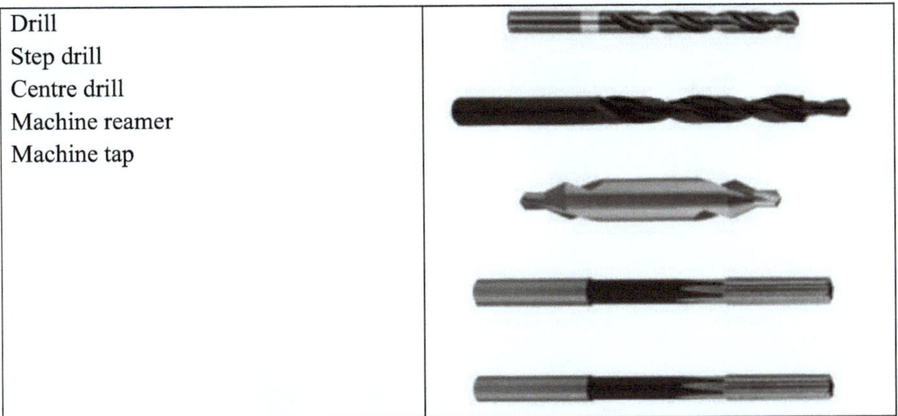 |

# Appendix 2: Types of inserts

**1-  Shooting**

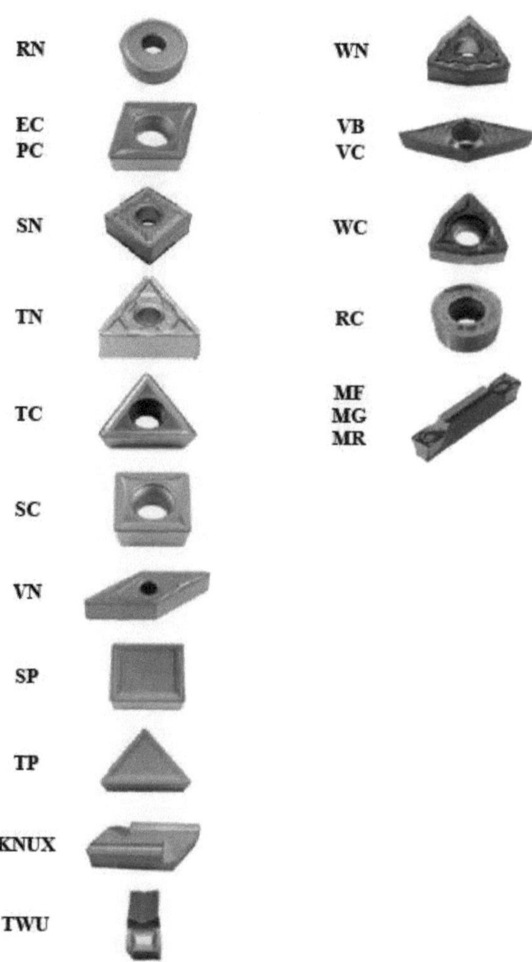

## 2- Milling

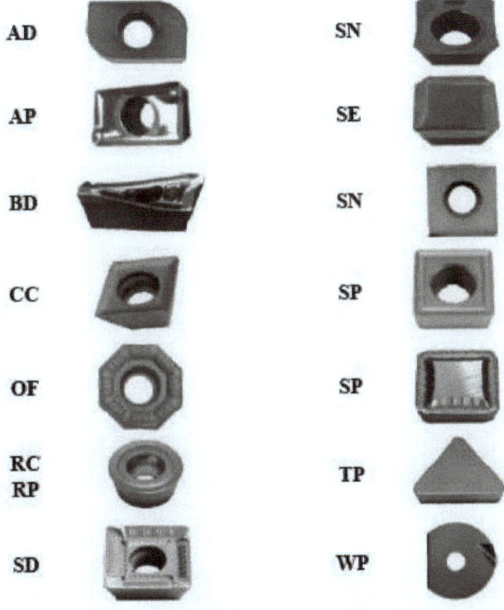

## I want morebooks!

Buy your books fast and straightforward online - at one of world's fastest growing online book stores! Environmentally sound due to Print-on-Demand technologies.

Buy your books online at
**www.morebooks.shop**

Kaufen Sie Ihre Bücher schnell und unkompliziert online – auf einer der am schnellsten wachsenden Buchhandelsplattformen weltweit! Dank Print-On-Demand umwelt- und ressourcenschonend produzi ert.

Bücher schneller online kaufen
**www.morebooks.shop**

info@omniscriptum.com
www.omniscriptum.com

Printed by Books on Demand GmbH, Norderstedt / Germany